2015
SHANGHAI
URBAN
SPACE
ART SEASON
MAIN
EXHIBITION
2015上海
城市空间艺术季
主展览

城市更新
URBAN REGENERATION
CAUP·IUAV
曹杨新村
社区公共空间城市更新
Prof. Weiqiang WANG
CAUP. TONGJI UNIVERSITY
Prof. Lan WANG
CAUP. TONGJI UNIVERSITY
Prof. Shiwen SUN
CAUP. TONGJI UNIVERSITY
Prof. Margherita Turvani
Prof. Maria Chiara Tosi
Prof. Stefano Munarin
Yujie XUE
CAUP. TONGJI UNIVERSITY
Xiji JIANG
CAUP. TONGJI UNIVERSITY
Alex Pelizer
Silvia Maroso

SUSAS
2015 上海城市
议程及活动

外滩滨水区
Waterfront Area of the Bund

上海历史文化风貌区保护规划
新江湾城
New Jiangwan City
多伦路风貌保护道路
Duo Lun Road Conservation Pathway

Hotel Sotelia
Enota
2007
SM House
Mimarlar Tasarim
Canakkale, Turkey
23
2007
WINNER
MUSAC - Contemporary Art Museum
of Castilla y León
2007
22
VM Housing
2007
20
Kastrup Seabath
2007
21
2007
19

2007
FINALIST
Sines Art Centre
Aires Mateus e Associados
Sines, Portugal
09
2007
2007 15
GlaxoSmithKline Day Nursery
Antonio Citterio & Partners
FINALIST
National Choreographic Centre
Rudy Ricciotti Architecte
Aix en Provence, France
2007
14
FINALIST
Mercedes-Benz Mus
UNStudio
Stuttgart, Germany
07
2007
2007
EMERGING ARCHITECT WINNER
Department of Mathematics
16 2007
Agbar Tower
Ateliers Jean Nouvel /
B720 Arquitectos
Barcelona, Spain
13
2007
Simone de Beauvoir Footbridge
Feichtinger Architectes
Paris, France
10
FINALIST
Phæno Science Centre
Zaha Hadid Architects
Wolfsburg, Germany
2007
06 20
2007 17
12
2007
Ecobulevard of Vallecas
Ecosistema urbano arquitectos
Madrid, Spain
11 2007
De Brug
JHK Architecten
Rotterdam, The Netherlands
Intervention on
and Surroundir
Antonio Jiméne
Granada, Spain

上海城市空间艺术季展览画册编委会 | SUSAS PUBLICATION EDITORIAL BOARD

# 2015 上海城市空间艺术季

# Shanghai Urban Space Art Season

## Main Exhibition

主办：
上海市城市雕塑委员会
承办：
上海市规划和国土资源管理局
上海市文化广播影视管理局
上海市徐汇区人民政府
Host：
Shanghai Sculpture Committee
Organizer：
Shanghai Municipal Bureau of Planning and Land Resources
Shanghai Municipal Administration of Culture, Radio, Film & TV
People's Government of Xuhui District, Shanghai

上海城市空间艺术季展览画册编委会 编
Edited by SUSAS Publication Editorial Board

# CONTENTS

# 目 录

# 序言

**上海市规划和国土资源管理局**
**上海市文化广播影视管理局**
**上海市徐汇区人民政府**

上海经过二十多年的快速发展，城市面貌焕然一新，市民生活明显改善，城市经济和活力令人瞩目。然而，与世界上很多大城市一样，由于人口加速集聚和城市快速扩张，环境、交通、安全等大都市的通病日益凸显，城市发展已经到了一个关键的历史性时期。上海城市面临的困难和挑战，城市未来的发展方向和发展方式，迫切需要得到社会各界的关注和积极参与，群策群力、众智众规。为此，上海市规划和国土资源管理局策划并发起两年一届的"上海城市空间艺术季"活动，希望打造一个以展示上海城市空间为平台的品牌活动，不断宣传和推广城市发展理念，促进城市空间塑造和公共艺术实践，展示城市优秀作品，呈现文化大都市魅力。

"上海城市空间艺术季"以"文化兴市，艺术建城"为理念。首届活动在上海市规划和国土资源管理局、上海市文化广播影视管理局、上海市徐汇区人民政府共同承办下，于 2015 年 9 月 29 日开幕，主题是"城市更新"，并通过展览与实践相结合的方式，将城市建设中的实践项目引入展览，将展览成果应用于城市建设的实践项目中。主展览设在徐汇滨江西岸艺术中心，通过历史传承、创新畅想、乡村乡愁、公共艺术四个角度诠释"城市更新"这一主题，将国内外先进的城市更新理论和实践全方位地进行展示和宣传。特别值得一提的是，不同于一般意义上的双年展，为加强实践性，本次空间艺术季活动在各区县设立了十五个实践案例展，选择与百姓生活密切相关的公共空间，如传统街区、工业遗产、市政设施、绿化广场、社区空间等，通过国际策展与公开征集等多种方式，展示城市空间公共艺术改造方案及实施效果。同时，将上海城市雕塑、公共艺术等相关展示活动及与市民生活密切相关的文化活动纳入形成联合展，使空间艺术季活动能不断地传播城市热点，美化城市空间，激发城市活力，提升城市魅力。

《2015 上海城市空间艺术季》出版物是本届空间艺术季活动的重要成果之一，共有两册，一册为主展览部分，包含主展览中"一条主线、四个维度、七个板块"的精华内容，一册为实践案例部分，包含所有十五个案例的空间创作和实践成果。这些丰富、详实的内容，希望能为关心上海城市更新工作的人们提供有益的借鉴，同时也期待更多市民参与城市空间品质提升这一有意义的工作。

# Preface

**Shanghai Municipal Bureau of Planning and Land Resources**
**Shanghai Municipal Administration of Culture, Radio, Film & TV**
**People's Government of Xuhui District, Shanghai**

Shanghai has taken on a new look with significantly improved life quality of its citizens and striking economy and vitality through more than 20 years of rapid development. Similar to other metropolises around the world, the urban development has entered a critical period when increasingly obvious problems appear in environment, traffic, safety, etc., due to rapid population increase and urban expansion.Attention and active engagement of all sectors of the society and wisdom and efforts of everyone are required to address the difficulties and challenges that Shanghai faces in determining urban development direction and pattern. To this end, Shanghai Municipal Bureau of Planning and Land Resources has planned and launched the biennial "Shanghai Urban Space Art Season" in the hope of building a brand event to showcase Shanghai urban spaces, communicate and promote the urban development concept, boost site projects in urban space shaping and public artworks, display excellent works in urban design and present the charm of the cultural metropolis.

Under the principle of "Culture Enriches City, Art Enlightens Space", Shanghai Urban Space Art Season (SUSAS) emphasizes on "International, Public, and Practical", thus to improve the quality of public space and add the city with glamour. It's hosted by the Shanghai Sculpture Committee and organized by the Shanghai Municipal Bureau of Planning and Land Resources, the Shanghai Municipal Administration of Culture, Radio, Film & TV and the People's Government of Xuhui District. SUSAS plays a key role in promoting the transformation of Shanghai, improving the quality of public space and boosting the urban regeneration. The event took "Expo + Site projects" model in order to promote a two-way exchange: introducing real cases from urban construction into the expo and applying the exhibits in practical projects. Located in West Bund Art Center in Xuhui Riverfront, The main exhibition interprets the theme of "Urban Regeneration" from four perspectives, namely Historical Inheritance, Innovation Creativities, Rural Nostalgia, and Public Art. Addition, different from other biennales, the SUSAS put on fifteen real cases of urban public space most closely related to the lives of residents, such as traditional district, industrial heritage, municipal amenities, green space and square, community space and earth art with a focus on how art-led-regeneration can transform urban public space. Meanwhile, we will incorporate relevant shows of Shanghai urban sculptures, public arts, etc. and cultural activities closely related to the life of citizens to form extensive exhibitions that continuously spread urban hot topics, beautify urban spaces, and enhance vitality and charm of the city.

*2015 Shanghai Urban Space Art Season* is one of the important fruits of the event. The book consists of two volumes, one describing the main exhibitions and its essence of "one main line, four dimensions and seven sections", and the other presenting 15 site projects including space works and practical fruits. We hope that the rich and detailed contents will provide useful reference for people who are concerned about the urban renewal of Shanghai and are looking forward to more citizens' engagement in the meaningful work of improving the quality of urban spaces.

# 上海城市更新的新探索

## 在"世界城市日"学术研讨会上的讲话

庄少勤

上海城市规划委员会办公室主任
上海市规划和国土资源管理局局长

城市更新伴随城市发展的全过程，折射着人们生活方式和城市发展方式的变化，是城市持续发展和繁荣的驱动者。不同地区、不同阶段的城市更新，呈现着不同的特点。

上海从千年之前东海之滨的小渔村，到七百多年前江南水乡的新县城；从1843年开埠，到1990年的浦东开发，乃至现在，一直处于城市更新的过程中。

与世界很多大城市一样，我们也经历了城市快速扩张、人口剧增等阶段，面临着旧城老化、服务能力不足等困扰。在这过程中，我们进行了以大规模旧区改造为代表的更新实践。但随着城市日趋长高、长大，以往"大拆大建"外延式扩张的发展老路已难以为继，注重提升城市品质和活力的内涵式发展成为当务之急。

### I 上海城市更新进入了一个新阶段

#### 挑战

上海的城市发展取得了举世瞩目的成就，但在土地利用、人口结构、空间品质、功能活力、文化传承、城市安全等方面也都遇到新的挑战。

一是土地利用。截至2014年底，全市建成区面积3124平方公里，超过市域陆地面积的45%，已逼近规划规模3226平方公里。与此同时，土地利用结构不够合理，工业用地比重过大，达27%，而公共设施和绿地的用地比例偏低；

二是人口结构。本世纪以来，上海人口激增800万，从2000年1608万人至2014年底已达到2425万，人口老龄化、少子化特征日益明显；

三是功能活力。上海不仅在保持传统工业优势面临较大挑战，而且在互联网等新科技和新经济背景下，传统服务经济的发展也面临较大压力，创新经济发展任重道远；

四是空间品质。城市游憩空间不足，人均公共绿地面积7.1平方米。养老及社区文化、体育等公共服务设施也相对不足；

五是文化传承。以往拆除重建为主的旧区改造方式，使上海城市历史风貌受到冲击。我们在延续历史文脉、留存城市乡愁方面要更加努力；

六是城市安全。上海也不时出现看海模式，全球气候变化异常等不确定性因素增加，对高密度、超大城市的安全和应急避难体系提出了更高要求。

**目标**

上海的国家使命是“当好改革开放的排头兵和创新发展的先行者”，并要在建设“国际经济、金融、贸易、航运等四个中心”的基础上，建设具有全球影响力的科技创新中心。

上海在新一轮城市总体规划中的愿景是建设一座追求卓越的全球城市。着力从城市竞争力、可持续发展能力、城市魅力三个维度，打造更加开放的创新之城、更加绿色的生态之城、更加幸福的人文之城。

**转型**

2014年5月6日上海召开第六次规划土地工作会议，韩正书记明确提出了“上海规划建设用地规模要实现负增长”，杨雄市长要求必须“通过土地利用方式转变来倒逼城市转型发展”，这标志着上海进入了更加注重品质和活力的“逆生长”发展模式。

新的城市发展模式要求城市治理机制创新，必须探索一条城市更新的新路。着力在空间上，打造一座更有安全感、归属感、成就感和幸福感的全球城市。

## II　上海城市更新的新理念

**更新理念变迁**

城市更新概念起源于西方，从“城市再开发”到“城市再生与城市复兴”；现代城市规划理论则从“现代主义”到“新城市主义”再到“生态都市主义”；甚至与东方的传统思想不谋而合。如1992年奥运会前，西班牙巴塞罗那借鉴中医理论，在城市更新中采取“针灸式”疗法，激发城市活力，成为东西文化相互交融下的经典更新案例。

中国先贤们的世界观强调“天人合一”，将城市看作天地间的有机生命体。在城市发展上遵循“以人为本，道法自然”，在城市规划建设中提倡“有之以为利，无之以为用”，讲究“因天材，就地利”，强调城市整体性、协调性和持续性，是对空间系统的辩证的、有智慧的处理方法。其实，西方城市规划先驱帕特里克·盖迪斯等学者也都提出过必须将城市作为活的有机体的理念，不过被强大的工业化边缘化了。而经历过工业文明的洗礼后，东西方殊途同归，形成这样的共识：城市发展应当回归生态文明，回归到亚里士多德时代的理想——城市让生活更美好。

**上海理念**

上海的城市更新正是吸取了东西方文明的成果，形成了中西结合、具有上海特色的“城市有机更新”理念。

上海的城市有机更新不仅将城市作为有机生命体，也是将大城市作为若干“小城市”的共生群体；不仅将城市更新作为城市新陈代谢的成长过程，也将城市更新作为一种对城市短板的修补和社会的治理过程；不仅强调历史人文和自然生态的传承，也强调城市品质和功能的创造；不仅是城市发展质量和效益提升的过程，也是城市各方共建、共治、共享的过程。

**上海特点**

在城市“逆生长”的模式下，上海有机更新有以下特点：

一是更加关注空间重构与社区激活，把社区作为一个功能完备的“小城市”。构建以社区为基础“网络化、多中心、组团式、集约型”的城乡空间格局；

二是更加关注生活方式和公共空间品质，强调以人为本，围绕社区构建生活圈，增强公共空间的品质和人性化的场所体验；

三是更加关注功能复合与空间活力，改变工业文明机械式的区划分割的做法。适应创新经济时代需求，围绕人的创新创业活动，通过土地混合使用，打造功能合理复合的创新空间，激发城市产业活力；

四是更加关注历史传承与魅力塑造，突出城市特色，提升城市魅力，营造出兼具历史底蕴和现代气质的城市文化禀性。上海目前正在开展成片保护历史街区计划，如田子坊，单从建筑本身没有太大保护价值，但这个街区代表的城市肌理和市民的生活方式，是城市生命的有机组成部分；

五是更加关注公众参与和社会治理，城市更新不仅是空间重组过程，也是利益重新分配的过程。应发挥市民的主体作用，注重社会多元协同（包括规划者、建设者、运行者、管理者和需求者），构建和谐有序、共建、共治、共享的社会关系；

六是更加强调低影响与微治理，注重以"小规模、低影响、渐进式、适应性"为特征的"中医式疗法"更新方式，推动城市的内涵式创新发展。

## III　上海城市更新的新方法和新实践

在学习总结多年实践经验基础上，上海市政府颁布了《上海市城市更新实施办法》，强调上海实施动态、可持续的有机更新，并注重以下工作原则：

一是政府引导，规划引领，政府制定更新计划，以区域评估为抓手，落实整体更新的要求，发挥规划的引领作用；

二是注重品质，公共优先。坚持以人为本，以提升城市品质和功能为核心，优先保障公共要素，改善人居环境；

三是多方参与，共建共享，创新政策机制，引导多元主体共同参与，实现多方共赢；

四是依法规范，动态治理，以土地合同管理为平台，实施全要素全生命周期管理，确保更新目标的有效实现。

**发挥文化引领作用，提升城市内涵。**

美国知名学者刘易斯·芒福德曾说"城市是文化的容器。城市根本功能在于文化积累、文化创新，在于传承文化，教育人民。"这段话精辟地阐述了文化、城市与人之间的关系。文化是城市的灵魂，发挥文化的引领作用，是对上海城市有机更新的基本要求和首要任务。

在美丽的城市一定遇见美好的市民。其实，对市民而言，城市就是一个大众创作、大众享有的公共艺术品。市民不仅仅是城市文化的被动接受者，更是积极创造者。城市更新就是大众艺术创作。将艺术注入城市空间，用文化来引领城市更新，不仅可以提升城市品质，更可以提高市民品味乃至修养品行。正是基于这样的思考，我们以“文化兴市、艺术建城”为理念，在今年举办了首届“上海城市空间艺术季”。

不同于一般意义上的双年展，上海城市空间艺术季不仅是面向专业人士，更是渗透到城市日常生活的每个角落，特别强调公众性和实践性。除主展览外，为进一步发挥街镇社区和市民作用，设立了十五个实践案例展，动态展示现实空间更新过程和效果。这也要求规划设计人员不仅要有人文情怀，还要有群众观念和实践能力。

**发挥社区平台作用，完善生活圈功能品质。**

以社区为基本生活单元，打造生活圈。首先要以市民需求和社区文体为导向，对更新地区进行综合评估，重点关注社区公共开放空间、公共服务设施、住房保障、产业功能、历史风貌保护、生态环境、慢行系统、城市基础设施和社区安全等方面内容，明确生活圈中“缺什么”，“补什么”，提供更加宜人的社区生活方式。

**发挥市民主体作用，促进城市共享发展。**

坚持以民为本，保障市民权益。探索“政府 - 市场 - 市民 - 社团”四方协同的机制，注重物业权利人和设计师及政府部门的协作，发挥市民协商自治作用。努力避免将城市更新变成加重社会两极分化的过程。

**发挥市场驱动作用，促进城市创新发展。**

上海的城市发展转型必须坚持减少对房地产的依赖。城市更新坚持公共利益优先，不以地块就地资金平衡为前提。不能让市场太任性，否则容易产生不公等问题；但要发挥好市场的资金、资源和创新能力等方面的作用。制定容积率奖励等方面的激励政策，多通过市场的方式，加强历史风貌保护、增加公共空间与改善公共服务。

**以契约管理为抓手，创新城市治理机制。**

每一个更新项目实施就是一次城市治理行动。而持续有效的治理取决于参与主体持续、稳定的社会责任。通过对项目主体社会责任的全生命周期契约管理，如将物业持有比例等要求纳入合同管理，减少投机因素，使开发商转型为城市运营商，与城市共同成长。这样的社会契约关联可改善社会治理机制，从源头上减少城市病产生，促进城市共建、共治、共享。

## Ⅳ　对未来上海城市更新的工作展望

城市更新已成为上海城市发展的主要方式，也是未来城市治理的关键抓手。我们将围绕“四个全面”的国家战略部署，主动适应和引领城市发展的新常态；根据“五位一体”的发展要求，不断探索“逆生长”模式下“有机更新”的新领域，持续提升城市品质和活力。

一是注重人本化。城市品质根本上取决于市民的品质，城市更新同样需要大众创新，通过人的更新成长来推动城市更新。我们将总结首届上海城市空间艺术季的经验，持续推进这一项活动的开展，以及正在编制的上海市民城市读本，SEA-Hi 城市空间艺术跨界论坛，“行走上海”等一系列活动，让公众参与城市更新，为城市规划工作搭建平台；

二是注重社会化。进一步完善社区规划体系与社区规划师制度，强化利益机制，如空间权益调节和激励机制。引导市场、社会主体和专业力量的积极参与，探索共享城市的治理模式；

三是注重信息化。积极应对智慧城市时代要求，建立动态的城市体征指标评估体系，构建更加开放的网络化城市共享共治平台；

四是注重法治化。在进一步实践基础上，适时提请市人大修订上海市城乡规划条例、上海市历史风貌保护条例，并进一步完善配套政策。

# Exploration on Shanghai Urban Regeneration

## Speech for Academic Seminar of World Cities Day

**Zhuang Shaoqin**
**Office Director of Shanghai Urban Planning Committee and President of Shanghai Planning and Land Resource Administration Bureau**

Urban regeneration accompanies the whole process of urban development, reflects changes in lifestyle and development mode of a city and drives the sustainable development and prosperity of a city. However, the features of urban regeneration vary from area to area, period to period.

From a humble fishing village by East Sea nearly one thousand years ago to a fledging town in Yangtze River Delta around seven hundred years ago, from the port opening in 1843 to Pudong Development in 1990, up to now, Shanghai never stops its regeneration.

With many of the metropolis around the world, Shanghai shares the experience of rapid expansion and population surge, as well as the problems of obsolete old towns and poor public services. We have practiced the regeneration featured by massive renovations. However, the city has grown out of the size for the old expansion and development mode featured by "mass-demolishing & mass-construction" and the connotation-based development that focuses on city quality and vigor improvement has grown high on the agenda for Shanghai.

### I The Shanghai Urban Regeneration has entered a New Stage.

#### Challenges

Despite the glaring achievements, Shanghai still confronts new challenges on land exploitation, population structure, space quality, function and vitality, cultural heritage and city safety.

Firstly, land exploitation. By the end of 2014, Shanghai has 3,124 $km^2$ of built-up area, exceeding the 45% of urban or land area and approaching the plan area of 3,226 $km^2$. Moreover, Shanghai is plagued by an improper land exploitation structure, excessive industrial land proportion (27%) and insufficient coverage of public utilities and green land.

Secondly, demographic structure. Since the beginning of the century, the population in Shanghai surged from 16.08 million in 2000 to 24.25 million by the end of 2014, up by about 8 million, increasingly characterized by population aging and low birth rate.

Thirdly, function vitality. Confronting profound challenges on maintaining the edge in conventional industrial and the pressure to develop conventional economic services against the backdrop featured by such new technology and new economy as the Internet, Shanghai has have a long way to go to innovate economic development.

Fourthly, space quality. With only 7.1 $m^2$ of public green land per capita, the public recreational space for Shanghai is under-developed. So it is for the elderly caring, community culture and sport & public service facilities in Shanghai.

Fifthly, cultural inheritance. The previous renovation dominated by removal and reconstruction dents the historic charm of Shanghai. We should ramp up the conservation of cultural heritage and local nostalgia.

Sixthly, city safety. Stricken by water-logging from time to time, Shanghai, confronting increasingly changeable global climate, should call upon higher requirements for safe and emergency asylum systems in high-density metropolis.

**Targets**

The mission of Shanghai is to "pioneer the opening up and innovation development" and to become a technical innovation center with international influence based on the role of "international economic center, financial center, trade center and navigation center".

In the new overall urban planning, Shanghai has the vision to grow into an international metropolis striving for excellence and is committed to build a more open city of innovation, a greener environment-friendly city and a happier culture-oriented city by city competitiveness, sustainable development capacity and city charm.

### Transformation

On May 6, 2014, Shanghai held the 6th Land Planning Meeting, where Han Zheng, General Secretary proposed to "de-grow the construction land in Shanghai" and Yang Xiong, Mayor of Shanghai requested to "transform Shanghai by transforming land exploitation". It marks the dawn of "de-growth" development mode of Shanghai, where attention is paid to quality and vitality.

As the new urban development mode requires for innovative urban governance mechanism, we should seek for a new approach for urban regeneration and strive to build an international city boasting stronger sense of safe, belonging, fulfillment and happiness.

## II New Philosophy of Shanghai Urban Regeneration

### Changes in regeneration philosophy

From "Urban Redevelopment" to "Urban Revitalization and Urban Rehabilitation", the idea of urban regeneration is originated from the West. Ranging from Modernism, New Urbanism and to Ecological Urbanism, modern urban planning theories even converge with Eastern traditional philosophies. For example, before 1992 Olympic Games, Barcelona, inspired by Chinese medicine theories, performed "an acupuncture therapy" to urban regeneration to vitalize the city, a classic case of urban regeneration that reflects the culture fusion between the West and the East.

The ancient sages of China upheld a world outlook that underlines "the Unity of Men and Nature" and regards a city as a living organism between the heaven and the earth. It is a dialectical and wise working method of space system to develop a city based on human-oriented approaches that follow the nature, resource availability and natural peculiarities and underline the globality, coordination and consistency of a city. In fact, a group of Western scholars, including Patrick Geddes, proposed the similar philosophy long time ago, i.e., to treat a city as a living organism. However, it was marginalized by the overwhelming trend of industrialization. After industrial civilization, the West and the East join hands again and reach the consensus that urban development should return to ecological conservation and the dream dating back to the times of Aristotle: Better City, Better Life.

**Shanghai Philosophy**

Shanghai draws lessons from both the Western and Eastern cultures to formulate the "Organic Urban Regeneration" philosophy with Shanghai characteristics.

For organic regeneration of Shanghai, cities are living organisms, and each large city is a mutualistic symbiosis group. Urban regeneration is not only a metabolic process, but also a process to eliminate shortfalls and govern society. Focus should be paid to historic and cultural heritage and natural bestowment as well as city quality and function. The organic regeneration means not only enhancing urban development quality and efficiency, but also sharing construction, governance and resources.

**Shanghai Characteristics**

With the mode of urban "de-growth", the Shanghai organic regeneration has the following features:

1. More attention is paid to spatial reconstruction and community activation. Each community is regarded as a fully functional "small city". We will build a community-based urban-rural layout that is "network oriented, multi-centered, clustered and intensive".

2. More attention is paid to life style, public space quality and human-centered philosophy. Community-based life sphere is built to improve the

public space quality and user experience of humane spaces.

3. More attention is paid to multi-functioning and spatial vitality to change the mechanical division characterized by industrial civilization. Adaption is made based on innovation and entrepreneurship to fulfill current economic requirements for creation, so as to build innovative spaces boasting rational and integrated functions and vitalize urban industries by mixed land exploitation.

4. More attention is paid to historic heritage and charm cultivation to highlight the characteristics and charm of Shanghai and develop a unique city identity that brings together historic heritage and modern radiance. Shanghai is launching a plan to protect extensive historic blocks such as Tianzi Lane. Despite the limited protection value for buildings, this block represents the city texture and life style integral to Shanghai.

5. More attention is paid to public participation and social governance. Urban regeneration is not only a spatial re-organization but also an interest redistribution. Citizens should play their major roles and focus on multilateral social coordination (including planners, constructors, operators, regulators and demanders) to create a harmonious, jointly built, corporately governed and shared society.

6. More attention is paid to minimize influence and subtle governance to achieve an urban regeneration characterized by "traditional Chinese medicine" thinking that boasts "small scale, minor influence, incremental progress and adaptation" to drive the connotative innovation development of Shanghai.

**III New Methods and Practices for Shanghai Urban Regeneration**

Based on years of practices and experience, Shanghai Municipal Government issued *Regulations on Implementation of Shanghai Urban Regeneration* to emphasize that Shanghai would perform a dynamic and sustainable regeneration according to the following principles:

Firstly, government instruction and planned guidance. Resorting to regional evaluation, the government plans regeneration to address the requirements for overall regeneration and guide the work as planned.

Secondly, quality counts and public first. Stick to the human-oriented approach and center on city quality and functions to safeguard public momentums and improve living environment.

Thirdly, multilateral participation and joint construction. Innovate policy-making system to guide multilateral participation and achieve win-win situation for stakeholders.

Fourthly, legal regulation and dynamic governance. Use land contract management as a platform to control whole life cycle for all factors to ensure effective fulfillment of regeneration targets.

**Enhance the connotation of Shanghai by cultural guidance.**

The famous US scholar Lewis Mumford said "a city is the container of culture, of which the fundamental function is to accumulate culture, innovate culture, communicate culture and to educate people". This concisely illustrates the relation between culture, city and people. As culture is the soul of a city, culture development is a fundamental requirement and overriding task for the organic regeneration of Shanghai.

You will see nice citizens in a beautiful city. For local citizens, a city is an artwork created and shared by the public. Citizens are the creators of urban culture, rather than only receivers. Urban regeneration is the artwork created by the public.It helps to improve city quality, citizen tastes and accomplishment to inject art into urban spaces and guide urban regeneration with culture.

Therefore, we launched the first "Shanghai Urban Space Art Season" with the philosophy of "Rejuvenating a City with Culture and Building a City with Art" from September to December 2015.

Different from other biennales, Shanghai Urban Space Art Season addresses to the daily lives of common people in Shanghai rather than only those of the professionals and elites and emphasizes public participation and practice. Except for major exhibitions, 15 practice cases were presented to show how an actual space is regenerated and what it is ended up with, so as to give full play to the roles of local communities and citizens. It calls for the humanistic sensibilities, public care and practice capability of planners and designers.

**Improve the living sphere with a community platform.**

We build a living sphere based on communities, the fundamental living units. First of all, the area to be regenerated should be comprehensively evaluated based on citizens' needs and community culture. Special attention should be paid to such factors as the public open spaces, utility facilities, industrial functions, relic protection, ecology & environment, slow traffic systems, urban infrastructure and community security of a community to "define and offset the community shortfalls" and realize a more pleasant community life.

**Citizens are taken as the main body to promote the shared development of Shanghai.**

The people-oriented philosophy to protect the citizens' interests should be followed. We should develop the coordination of "government-market-citizens-associations" and underline the cooperation of property owners, designers and authorities to exert the autonomous negotiation of citizens. Escalation of social polarization caused by urban regeneration should be avoided.

**The driving role of the market is fully played to promote the innovation of urban development.**

The dependence on real estate industry of Shanghai urban transformation must be curbed. For urban regeneration, priority should be given to public interests rather than fund balance of the plot. We should control the market to avoid partiality and give scope to the fund, resource and creation of market. Such incentive policies as nonresidential density awards should be formulated to improve historic heritage protection, public space coverage and public service by market.

**Innovate the mechanism of urban governance by contract management.**

Every regeneration project means an urban governance action. Persistent and effective governance depends on the persistent and reliable social commitment of major participants. Whole-life-cycle contract management on the social commitment of major project participants, i.e., to include the property holding ratio into contract management, can reduce speculation factors and turn developers into urban operators and make them grow with the city. Such social contract link helps to improve social governance mechanism and eradicate urban illness to enhance the sharing of construction, governance and resources.

## IV Outlook for Shanghai Urban Regeneration

For Shanghai, urban regeneration has become a major growing approach and the key for future urban governance. Centered on the national strategic deployment of "Four Comprehensives" (comprehensively build a moderately prosperous society, comprehensively deepen reform, comprehensively implement the rule of law and comprehensively strengthen Party discipline), we should take initiative to adapt to and lead the new normal of urban development. Based on the requirements for "Five-in-one" development (economy, politics, culture, society and ecological conservation), we should keep exploring “Organic Regeneration” under “De-growth” mode and improving urban quality and vitality.

Firstly, people oriented: As the quality of a city depends on the quality of citizens, urban regeneration cannot do without public innovation. Urban regeneration should be promoted by the renewal and development of people. We are now formulating a city handbook for Shanghai citizens, running SEA-Hi forum, producing a documentary named "Walk in Shanghai" and encouraging citizens to participate urban design.

Secondly, society oriented: We improve the systems for community planning community planers and enhance interest control mechanism, such as space rights adjustment and incentive mechanism. Participation of market, social subjects and professionals were guided to explore the mode of shared urban governance.

Thirdly, information oriented: We address to the current requirements for smart city, build dynamic city evaluating index system and establish a more open network-based platform for joint building and corporate governance of city.

Fourthly, law oriented: Based on practices, we submit regulations of rural-urban planning and historic landscape protection of Shanghai to Municipal People's Congress for revision and improve the supporting policies.

# 城市更新与文化创新

**胡劲军**
**上海市文化广播影视管理局局长**

万众瞩目的上海城市空间艺术季落下帷幕了。本届城市空间艺术季以“城市更新”为主题，聚焦“空间”和“艺术”关键词，充分演绎了城市更新的主题理念，展现了实践城市更新的经典案例，汇聚了探索城市发展的智慧思考，实现了市民参与城市文化氛围营造的诸多可能。

在城市加快更新的大背景下，上海城市发展既面临难得的发展机遇，也面临严峻挑战。究竟如何更好更快地发展和繁荣城市，提升市民生活、工作的城市公共空间品质，切实增强市民的生活幸福感和文化获得感？这需要城市更新与文化创新同步规划设计、同步组织推动。2015 上海城市空间艺术季的举办，正是谋求城市更新、城市发展的一次重要尝试。这是一次以城市空间为载体，以文化艺术为内容，促进城市更新提升城市品质的创新实践。文化是城市的灵魂和命脉。城市是文化、社会和经济活动的摇篮。文化的语言是城市最通俗的语言。上海城市空间艺术季将文化艺术作为城市更新的重要引擎内容，贯穿于活动的全程、城市更新的始终。无论是演绎艺术季主题理念的主展览，展现城市更新多元实践的案例展、拓展艺术季内涵外延的联合展；还是激起智慧碰撞的主题论坛、激发参与热情的市民活动，艺术季以城市空间为载体，用文化活动激发了城市新的生机和活力，用艺术作品点亮了城市空间，营造了浓郁的城市公共文化氛围。城市空间艺术季的举办，首次打造了具有“国际性、公众性、实践性”的城市空间艺术品牌活动，成为发挥城市文化创造力、凝聚力、影响力，更好服务城市更新发展、提升城市内涵品质的一次创新实践。

这是一次以跨界融合为引擎，以创造更新为路径，建立城市更新统筹协调平台的合力之举。文化融合是世界文化发展的必然趋势。城市更新的过程是各种文化相互碰撞、激荡的过程，城市更新与文化创造的融合尤为必要。这有利于使各种文化和谐统一于城市更新中，形成合力，共同推动城市的整体发展。

本届城市空间艺术季以文化与规划的融合，文化、艺术、建筑、规划的跨界为特点，首次将艺术家、建筑师、策展人聚集在一起，创造了多个跨界融合的合力之举。在组织机构上，市区两级联动，文化与规划部门合作办节；在资源整合上，国际与国内专家联合策展；在平台联动上，上海城市空间艺术季与上海市民文化节首度联手，这一跨界融合、群策群力谋划城市更新的做法，在全国乃至全球范围内也尚属首次。

上海城市空间艺术季的举办，启动了跨界融合的引擎，吸引了众多跨界艺术家、多元社会主体，投入到上海城市更新和发展中来。并以文化创造与城市更新相结合为路径，为建立城市更新统筹协调平台，推进城市更新中文化的发展与繁荣提供了整体合力。

这是一次以城市更新为契机，以市民福祉为归宿，营造浓郁城市公共文化氛围的升级设计。处于城市更新中，人们的思想活跃、观念多元、行为多向，这要求城市的发展，要特别注重把文化放到更为突出的位置。既重视城市更新中，营造浓郁的城市公共文化氛围，满足市民多样化的文化需求，又要注重以市民福祉为中心，增强市民的文化获得感和主体性。

城市需要文化，文化需要氛围。2015 上海城市空间艺术季期间，透过上海市民文化节平台，持续推进美术、非遗、演艺、群文等各类展览、展示、展演，进地铁、进街头、进绿地、进商圈、进交通站点、进机场码头工程。在城市建筑、道路广场、轻轨车站、公园绿地等空间载体功能中，注入文化氛围因素，使文化无处不在、无处不见，广泛渗透于城市更新全过程。

城市是由生活其间的每个人构成的，市民不应该仅仅是文化的消费者、欣赏者，更应该是参与者、创造者和推进者。2015 上海城市空间艺术季、上海市民文化节首度联手，共同推出"100 个上海最美城市空间"和"100 个上海城市空间塑造案例"的征集、评选活动。活动正是以市民为主体，邀请市民"为上海点赞"、"为城市支招"，激发更多市民参与到城市空间的塑造、创新中。以城市更新为契机，促进城市公共文化氛围营造的升级设计，推动上海城市空间更好的为城市居民服务。

城市空间艺术季的举办，是促进城市更新提升城市品质的一次创新实践，是建立城市更新统筹协调平台的一次合力之举，是营造浓郁城市公共文化氛围的一次升级设计。这也仅仅是城市更新与文化创造有机融合，同步规划、合力共谋城市发展的一个起点。

# Urban Regeneration and Cultural Innovation

**Hu Jinjun**
**Director General of**
**Shanghai Municipal Administration of Culture, Radio, Film & TV**

The much-anticipated Shanghai Urban Space Art Season (SUSAS) has dropped its curtain. This Art Season is themed by 'urban regeneration' and focuses on 'space' and 'art'; the two key words fully express the thematic idea of urban regeneration, show typical cases of urban regeneration, gather wisdom and thoughts in regard of urban development, and realized citizens' participation in the city's culture construction.

In the process of urban regeneration , Shanghai is facing both opportunities and severe challenges. How to realize better and faster development of cities? How to improve the quality of urban living and working space, strengthen the citizens' life satisfaction and cultural acquisition? This requires a synchronized planning, design, organization and promotion of cultural innovation. The SUSAS 2015 is an important attempt in this direction.

This is an innovative practice based on urban space and focused on culture and art; it will promote urban regeneration and improve the quality of the city.

Culture is the soul and lifeline of a city. City is the cradle of cultural, social and economic activities. Cultural language is the plainest language of a city. SUSAS has taken culture and art as an important engine content of urban regeneration, and made it run through the whole activity and the urban regeneration process.

The main exhibition has conveyed the thematic concept of the Art Season; the case studies exhibition has showed diversified practices of urban regeneration; the joint exhibition has extended the connotation of the Art Season; the forum session has stirred up wisdom collision; and the citizen activities have inspired the citizens' enthusiasm in participation. Based on urban space, the Art Season has inspired the city's new vigor and vitality, lit up the urban space with art works, and created a rich urban public cultural atmosphere.

SUSAS has created the urban space art brand activities for the first time; the activities are international, public and practical. It has become an innovative practice that will utilize the creativity, cohesion and influence of urban culture as well as better serve the urban regeneration and development and improve the connotation quality of the city.

This is the fruit of cooperation; it is based on innovative renewal and uses cross-boundary fusion as the engine to establish the planning and coordination platform of urban regeneration.

Cultural integration is the inevitable development trend of world culture. The process of urban regeneration is a process that witnessed cultural collision and surge; that's why urban regeneration and fusion of cultural innovation is especially necessary. This will enable various kinds of cultures to unify harmoniously into the urban regeneration and form composition forces to jointly promote the overall development of the city.

SUSAS is characterized by the integration of culture and planning and the integration of culture, art, architecture and planning. It has gathered artists, architects and curators together to harvest the fruit of cooperation. In regard of organizational structure, the municipal level and the district level took joint action and the cultural departments and planning departments cooperated with each other; in regard of resources integration, international and domestic experts jointly curated the event; in regard of platform linkage, SUSAS cooperated with the Shanghai Citizen Cultural Festival for the first time. It's the first time in China even in the world to realize urban regeneration through such a cross-boundary fusion.

SUSAS has started the cross-border integration and attracted a lot of trans-boundary artists and pluralistic social main bodies to involve into the urban regeneration and development of Shanghai. It has also integrated cultural creation and urban regeneration to establish the planning and coordination platform of urban regeneration and provide overall joined forces to the cultural development and prosperity of urban regeneration.

We take urban regeneration as an opportunity to improve public welfare and create a richer cultural atmosphere in urban public space.

During the process of urban regeneration, people's thoughts are very active, concepts are diversified, conducts are multidirectional. This requires the urban development to focus on culture. When engaging in the urban regeneration, we should create a rich urban public cultural atmosphere, meet the diversified needs of citizens, and pay attention to public welfare and strengthen citizens' sense of cultural acquisition and ownership.

A city should create the right environment for culture development. During the SUSAS 2015, Shanghai continues to promote various kinds of exhibitions and demonstrations of arts, intangible cultural heritage resources, performing arts and folk arts. Relevant teams went to subways, streets, greenbelts, business circles, transport stations and airports and wharfs. The city has also injected cultural factors into spatial carriers such as urban buildings, roads, squares, light rail stations and park green spaces; this has enabled culture elements to be seen and felt everywhere and widely penetrated into the whole urban regeneration process.

A city relies on all the people living in it; citizens should not be merely consumers and appreciators, but should be participants, creators and impellers of culture activities. For the first time, SUSAS cooperated with the Shanghai Citizen Cultural Festival and jointly launched the solicitation and selecting activity of the "100 Most Beautiful City Spaces in Shanghai" and the "100 Shanghai Urban Space Establishment Cases". This campaign invited citizens to give their thumbs up for Shanghai and offer support, in this way inspired more people to participate in the building and improvement of the urban space. The Art Season has taken the urban regeneration as an opportunity to promote the upgrade design of the establishment of urban public cultural atmosphere and enable Shanghai's urban space to better serve the citizens.

SUSAS 2015 is a ceative practice to promote urban regeneration and improve the quality of the city; it is the fruit of cooperation to establish a coordination platform of urban regeneration; it is an upgrading design aiming at creating a rich urban public cultural atmosphere. What's more, it's also a starting point for the organic fusion of urban regeneration and cultural creation, synchronous planning and cooperated development.

# 助力文化传承，勇于开拓创新，顺应城市工作新形势

鲍炳章
徐汇区区长

上海城市发展已经进入新的阶段，如何借鉴发达国家城市的成功经验，改善城市人居环境，提高居民生活质量，丰富城市文化内涵，盘活土地存量而非增量，提升城市管理功能，实现城市公共服务设施、基础设施、文化设施的均衡供给，推动城市品质再上一个新阶梯，是上海城市发展面临的机遇与挑战，而城市有机更新将在文化创新的引领下成为应对这一机遇和挑战的有效方法。

徐汇区政府一直在不断地摸索和创新。在刚刚闭幕的上海城市空间艺术季中，主展览场馆就位于徐汇滨江的西岸艺术中心。对于徐汇滨江地区的开发，城市更新的重点在于土地的二次开发、用地性质和功能的转换、工业区转型以及滨水区的整治和改造，不"大拆大建"，一方面保留原有的工业建筑，加以更新改进成为保留城市工业记忆的新城区，例如工业塔吊、老船屋变身新的文化景观，一些航空油罐甚至成为极具创意的演艺空间，徐汇西岸艺术中心也是由上海飞机制造厂车间改造而成；另一方面，以现代文化艺术为背景打造"西岸文化走廊"品牌工程，着力提升城市公共开放空间品质，比如龙美术馆、余德耀美术馆、跑道公园等。同时，在滨江沿岸通过清除路障、加强安全配套设施、建立塑胶跑道等方式，艺术化地提升了街道空间品质，吸引了很多市民自主参与，推动了社区空间品质和活力的改善，把徐汇滨江打造为独具魅力的文化传媒产业集聚区和充满活力的滨水公共活动区。

除了主展览，本次艺术季中，我们还通过几个案例展，举办了内容丰富、形式多样的城市更新活动。这些活动将艺术注入到城市空间，用文化引领了城市更新。对于上海中心城区最大的风貌区——衡复历史文化风貌区，城市更新更应基于社区记忆和文化传承，着力弘扬海派城市文化。我们邀请陈丹燕女士做客城市更新论坛，口述社区记忆弘扬社区文化；举办尔冬强先生专题讲座，向公众解读 Art Deco 文化艺术；世界小学"老洋房探踪"特色拓展课程有力地显示了海派城市文化底蕴，为社区公共参与和建设提供了启示；风貌整治和老建筑改造设计交流研讨会从改进设计和实施模式等问题出发，分享了从建筑设计到管理和实施部门人员的交流，摸索出了相关切实有效历史风貌区建筑保护更新的做法；"徐汇风貌道路保护规划与实践 2007—2015"专业地展示了徐汇区在 2007 年以武康路为保护规划和整治实施试点到 2015 年全区风貌道路保护规划所取得的新思路和新模式。

在新的城市发展形势下，做好城市工作要抓住城市更新这条主线，继往开来，既要做好传统文化传承，又要勇于开拓创新，不断完善城市管理和服务，让居民在城市生活得更方便、更舒心、更美好。愿我们共同努力，建设好我们共同的城市。

# Boosting Cultural Heritage, Braving for Innovation and Adapting to the New Situation of Urban Work

**Bao Bingzhang**
**District Mayor of the People's Government of Xuhui District**

As Shanghai has entered a new development stage, it is now faced with opportunities and challenges in how to use the successful experience of cities in developed countries to improve urban residential environment, enhance residents' living conditions, enrich urban cultural connotation, make good use of land reserves (rather than land supply), promote urban management functions, achieve balanced supply of public facilities, infrastructures and cultural facilities in the city as well as push forward the further advancement of urban quality; the organic regeneration of the city will be an efficient method to respond to those opportunities and challenges under the direction of cultural innovation.

The People's Government of Xuhui District has been in constant exploration and innovation. In the just-concluded Shanghai Urban Space Art Season, the main exhibition venue is in West Bund Art Center, which locates in riverside of Xuhui District. For the development of riverside area in Xuhui District, the focal point of city regeneration is the secondary development of land, the conversion of the nature and function of land, the transformation of industrial area and the renovation & retrofit of riverside area; demolishment and reconstruction in large scale are to be prohibited, the original industrial buildings are kept and upgraded into the new area with the urban industrial memory on one hand such as the transfiguration of industrial tower cranes and old boathouses into new cultural landscapes; some aviation fuel tanks have even become innovative performance spaces. Xuhui District West Bund Art Center was renovated from the workshop of Shanghai Aircraft Manufacturing Factory. On the other hand, the branded project of "West Bund Culture Gallery" is to be built against the background of modern culture and art to improve the quality of urban public open spaces, such as Long Museum, Yuz Art Museum, Runway Park, etc. At the same time, the quality of street space has been improved artistically by clearing the roadblocks, enhancing the safety supporting facility, building plastic tracks, etc. along the riverside, a lot of citizens are attracted to participate voluntarily, in this way the improvement of quality and vigor of community space is boosted. The riverside area of Xuhui District has become a cultural media industry cluster district with unique charm and a waterfront of public activities that is full of energy.

During the Art Season, in addition to the main exhibition, we have also demonstrated a couple of case shows (cases in Xuhui Historic Area Regeneration 2015, Caohejing Development China Fortune Industrial Park, and special municipal exhibition of urban regeneration of Xuhui District that starts from overpasses) to hold urban regeneration events that is rich in various contents. Urban space is infused with art through these events, and the city regeneration has been led by culture. For the biggest historic area in downtown Shanghai - Hengfu Historic and Cultural Area, the urban regeneration shall be based further on the community memory and cultural heritage to propagate Shanghai urban culture. For this purpose we have taken a lot of explorative measures of protection and renewal, for example, we invited Chen Danyan to the City Regeneration Forum to tell about community memory and promote community culture; organized a special lecture for Er Dongqiang to interpret Art Deco culture and art to the public; Moreover, special curriculum Old Houses of World Primary School that vigorously demonstrated the cultural heritage of Shanghai urban culture and provided inspiration for the construction and participation of community public; feature renovation and design & reconstruction seminar derived from improvement design, implementation mode and other issues to share the experiences of personals from architectural design department to management and implementation department, developing a relevant and effective way for protection and renewal of historic area; Xuhui District Historic Road Conservation Planning and Practice 2007-2015 has demonstrated professionally the new ideas and new models acquired from the process from protective planning and implementation of renovation of Wukang Road as a trial in 2007 to the protective planning of the historic roads in the whole district in 2015.

In the new urban development situation, urban work shall follow closely the mainline of city regeneration; both the traditional cultural heritage and the bold exploration & innovation shall be emphasized; urban management and service shall be constantly improved to make city life more convenient, comfortable and happy. Let us work together to develop Shanghai into a better city.

# 城市更新是理想、艺术和价值的体现

**郑时龄**

**中国科学院院士**

**2015 上海城市空间艺术季**

**学术委员会主席**

2015 上海城市空间艺术季在上海市规划和国土资源管理局、上海市文化广播影视管理局、徐汇区人民政府的大力领导下，在策展团队将近一年的努力工作和各区县、街道、学校、企业和市民、艺术家、建筑师、规划师、政府官员的通力合作下如期举办，许多人为之付出了无法量度的辛勤劳动和心血。城市更新是城市理想、艺术和价值的体现。如同城市更新一样，城市空间艺术季是一个展示，更是一个培育理念和实践的过程，让艺术介入空间，进入生活，通过城市更新植入面向未来的理想。

2015 上海城市空间艺术季是上海城市更新、重视城市空间和城市艺术的里程碑。在城市发展的这个阶段，注意城市空间的品质，注重城市更新和城市空间的修补，维系城市的特色，使城市空间品质更为优秀、更为生态、更为宜人、更为宜居也更为宜业。

古希腊哲学家赫拉克利特说过：" 看不见的和谐比看得到的和谐更美好。" 城市更新就是努力去营造看得到的和看不到的和谐。城市更新既涉及物质性，也涉及非物质性的更新，包括城市空间结构的更新、建筑的更新、城市环境和道路的更新、城市意识的更新。城市更新如同公元前 1000 年古希腊神话中的英雄伊阿宋取金羊毛所乘坐的那艘船阿尔戈号，历尽千辛万苦，这艘船的每一块木板都随着时间而被更替，但它仍然是阿尔戈号，是一个有可识别意义的整体。那些优秀的历史城市在经过长期的更新后会更加辉煌。

在政治、历史、经济、意识形态、文化、宗教、伦理、人口等种种因素的作用下，城市总是处于更新过程中。即使欧洲那些中世纪和文艺复兴时期留存至今的城市，城市空间虽然没有经历剧变，然而城市依然在更新，内在的和外显的更新。城市更新有不同的规模，或者是累积添加的城市开发，大型的建设项目，或者是城市的蔓延，新的城市元素取代已经存在的元素，将消极的城市空间转换为积极的公共空间。城市更新包括新区开发、旧城区复兴、用地性质和功能的转换、工业区转型、城市美化、港区和滨水区的整治和改造，以及近年来的城市生态规划和可持续发展等。城市更新有着程度和规模的差异，既有拼贴式小修小补，也有如同 19 世纪巴黎奥斯曼计划那样的 " 一年一个样，三年大变样 " 大刀阔斧的翻新，取决于城市的理想和发展目标。

城市总是多元的、多样化的，且城市是在历史中形成的，城市需要有特色，富于艺术创意，而又丰富多彩的城市空间。城市空间是政治、经济、文化、宗教、艺术等因素所形成的，那些让我们永远仰慕的城市空间总是生活的空间，适合人们停留和交往的空间，总是与文化、社会融合的空间，总是有着创造性的空间。上海靠大规模建设城市空间的时代已经结束，城市的发展不再追求宏大叙事，追求大手笔，不是靠伟大，靠纪念性，或者靠奢华地堆砌材料和体量赢得人们的赞赏，而是塑造人们切实生活在这里的空间，能让人们在这里欣赏并愉快地停留，激发人们的创意，让人们更热爱生活，热爱我们的城市。需要我们不断地更新和修补城市空间，从而创造更美好的城市空间。

愿我们大家共同努力，让我们的城市更美好，使我们的生活更美好。

# Urban Regeneration: Ideal, Art and Value

**Zheng Shiling**
**Academician of the Chinese Academy of Sciences;**
**Chairman of SUSAS Academic Committee**

"Shanghai Urban Space Art Season 2015" (SUSAS 2015) has been held as scheduled under the strong leadership of Shanghai Municipal of Planning, Land Resources, Shanghai Municipal Administration of Culture, Radio, Film and Television, and the People's Government of Xuhui District and on the basis of one-year arduous efforts of the curatorial team and wholehearted cooperation among citizens, artists, architects, planners and government officials from various districts, counties, sub-districts, schools, enterprises. Many people have paid immeasurable hard work and efforts for this event to take place. Urban renewal is the manifestation of urban ideal, art and value. Just as urban renewal, SUSAS is a demonstration, and more importantly, a process of cultivating concepts and practices, to integrate art in spaces and life, incorporate future-oriented ideals through urban renewal.

SUSAS 2015 is a milestone in Shanghai's urban renewal with the emphasis on urban spaces and urban art. As the stage of urban development, we focus on the quality of urban spaces, value urban renewal and space renovation, and maintain distinctive features of a city, thus improving the quality of urban spaces to offer more ecological, agreeable, livable and industry-adaptable spaces.

Ancient Greek philosopher Heraclitus said, "The hidden harmony is better than the visible." Urban renewal is aimed at creating both visible and hidden harmony. It refers to renewal of the tangible and intangible aspects, including spatial structure, architecture, urban environment and roads as well as urban consciousness. It resembles the ship Argo taken by Greek mythological hero Jason to bring the Golden Fleece back around 1,000 B.C. Although the planks of the ship were replaced after undergoing many years of turmoil and hardship, it is still the Argo, a ship with recognizable meaning as a whole. Those outstanding cities with historical significance always become more glorious after long-term renewal.

Cities keep updating under the function of multiple factors such as politics, history, economy, ideology, culture, religion, ethics and population. As to the existing European cities traceable to the Medieval Age and the Renaissance, even if their urban space experiences no drastic transformation, the urban renewal is under way continuously in a covert and overt manner. Urban renewal boasts different scale, accompanied by cumulative urban development with large construction projects or urban sprawl with existing urban elements replaced by new ones, and negative urban space converted into positive public one. Urban renewal includes new zone development, old town revival, transformation of urban land properties and functions, transformation of industrial zone, urban ornament, harness and renovation of harbor and coastal areas as well as urban ecological planning and sustainable development in recent years, etc. Determined by the urban ideal and development goals, urban renewal differs in the scale to the degree with both collage-like tinkering and drastic bold renovation similar to "Minor Change Every Year and Great Change Every Three Years" described in "Haussmann Plan" of Paris in the 19th century.

Cities are always diversified and pluralistic. Moreover, they are formed in the history, so they need coloful urban space with characteristics and artistic creativity. The urban space comes into being out of such factors as politics, economy, culture, religion and art. The urban space arousing our perpetual admiration is always suitable for life and people's stopover and interaction, integrated with culture and society as well as creativity. The era has come to an end when Shanghai carried out large-scale construction of urban space. The city no longer develops in pursuit of grand narration or large scale, nor attempts to win people's admiration on the basis of grandeur, monumentality or luxuriantly stacking materials and volumes. Instead, it develops by shaping the space where people live in reality, enabling people to appreciate and stay merrily, inspiring their creativity and making them love their life and city more. It is necessary for us to continuously renew and renovate urban space to create a more beautiful urban space.

Let's make joint efforts to make our city more beautiful and our life better.

# 城市品质提升新路径：艺术介入城市公共空间

吴为山
中国国家美术馆馆长
2015 上海城市空间艺术季学术委员会主席

由上海市城市雕塑委员会主办，上海市规划和国土资源管理局、上海市文化广播影视管理局、上海市徐汇区人民政府共同承办的“城市更新——2015 上海城市空间艺术季”终于在金秋的九月与公众见面了。这是上海在促进城市转型发展中的又一宝贵尝试，也凸显出上海作为中国城市发展尖兵的敏锐与前沿。

在告别以往“大拆大建”式的城市发展模式之后，“城市更新”将是未来上海与中国城市发展的新特征。面对非增量发展模式的城市空间，要实现由物质向品质空间的转变，必须重视艺术这一引领与激发城市内涵的重要因素。我非常赞同“2015 年上海城市空间艺术季”提出的“空间艺术”，强调以艺术对城市公共空间的介入与互动，并此构建出城市从数量扩张到品质提升的城市发展崭新路径。事实上“城市更新”问题，不仅形式上要更新，更强调心理空间更新、理念更新，通过空间上的更新、形式上的更新，引领人们心理上文化重生。公共空间艺术由此与城市的基础设施一样，成为城市发展的必需品。艺术在与空间、建筑、景观的融合中，重新改善着城市空间与人的关联，展现地域文化和时代精神。

而我们在涉足这一实验领域之时，也要清楚地认识到，艺术与空间的结合与共谋势必要关联起更为广阔的文化内涵。而这种内涵应在保持自己文化特色、继承中华民族传统文化、吸收世界优秀文化的基础上，与所在城市特征紧密关联。上海城市空间艺术的更新，第一要有传统的精神，第二要有现代的视觉特征，第三要有创新的原动力，这三者融汇一体、多元共生才能把上海建设成国际的上海。上海的地域文化与整个中华文化洪流融渗，才能催生城市的国际竞争力。

作为此次展览的学术委员会主席以及参展艺术家，我很荣幸参与其中，见证了这一重要时刻。我也衷心希望上海能在众多专家学者、建筑师、规划师、艺术家的共同探讨与努力下，为中国新常态下的城市发展建设开辟出一条新路！也期待公共艺术与城市空间的相互滋养与碰撞下，生发出更为璀璨的艺术盛景！

# A New Path to Improve Urban Quality: Intervention of Art into Public Space

**Wu Weishan**
**Director of The National Art Museum of China；**
**Chairman of Academic Committee of SUSAS 2015**

*Urban Regeneration: Shanghai Urban Space Art Season 2015* hosted by the Shanghai Sculpture Society and co-organized by the Shanghai Planning and Land Resource Administration Bureau, Shanghai Municipal Administration of Culture, Radio, Film & TV and Shanghai Xuhui District People's Government kicked off this September. The event is another valuable attempt of Shanghai to boost urban transition, which underlines the posture of Shanghai, a farsighted and discerning pioneer in urban development.

Urban Regeneration has replaced the previous development mode of "mass-demolishing & mass-construction" to become the new development characteristics for Shanghai and other cities in China. For non-incrementally developing urban spaces, attention should be paid to art, a crucial factor that leads and inspires a city's spirit connotation so as to shift material construction to quality enhancement. I can't agree more with the idea of Artistic Space proposed in Shanghai Urban Space Art Season 2015, which stresses the intervention and interaction of art in urban public spaces to blueprint a new city development that turns quantity expansion to quality improvement. In fact, Urban Regeneration entails the update not only on style, but more on mental space and philosophy. By updating the space and style, a mentally cultural regeneration can be achieved. Therefore the public space art will, like urban infrastructure, become a necessity for city development. By merging spaces, buildings and landscapes, art improves the tie between urban spaces and people and manifests local culture and spirit of the time.

When exploring such an experimental field, we should understand that the combination and coordination of art and space are bound to involve a more extensive cultural connotation which should be closely linked to the city characteristics based on maintaining the cultural features, inheriting Chinese cultural heritage and absorbing fine cultures around the world. Three factors are essential to update the urban space art in Shanghai: traditional spirit, modern visual features and driving force of innovation. Only by combining those three factors and making them a multiplex symbiosis can we build

Shanghai into a strong international metropolis. Only by converging Shanghai's local culture to Chinese culture, can we stimulate the international competitiveness of Shanghai.

I am much honored to attend this event and witness this momentous moment as the Chairman of Academic Committee and an artist. Hopefully, Shanghai, based on the exploration and efforts made by scholars, architects, planners and artists, can forge a new path for city development and construction under China's new normal. We expect the complement and collision of public art and urban space will bring about a more profound art boom!

# 组织架构

# ORGANIZATION STRUCTURE

## 执行团队 | SUSAS EXECUTION TEAM

**在上海市城市雕塑委员会统筹指导下**
**推进艺术季具体工作开展**
**To Work as Coordinated and Instructed by**
**Shanghai Urban Sculpture Commission**

--------------------

### 俞斯佳 | Yu Sijia

上海市规划国土资源局总工
上海市城雕办副主任
Chief Engineer of Shanghai Municipal Bureau of Planning
and Land Resources;
Vice Director of Shanghai Urban Sculpture Planning
and Management Office

### 滕俊杰 | Teng Junjie

上海市文化广播影视局艺术总监
Art Director of Shanghai Municipal Administration of
Culture, Radio, Film & TV

### 徐建 | Xu Jian

徐汇区副区长
Vice District Mayor of the People's Government of Xuhui District

**上海市规划国土资源局风貌处**
**上海市文化广播影视局艺术处**
**上海市城市公共空间设计促进中心**
**徐汇区规土局**
**上海西岸开发（集团）有限公司**
**上海市公安局、财政局、建管委、绿化市容局、旅游局相关处室**
**Landscape Division, Shanghai Planning and**
**Land Resource Administration Bureau;**
**Art Division, Shanghai Municipal Administration**
**of Culture, Radio, Film & TV;**
**Shanghai Design & Promotion Center for**
**Urban Public Space;**
**Planning and Land Resource Administration Bureau**
**of Xuhui District, Shanghai;**
**Shanghai Xi'an Development (Group) Co., Ltd.;**
**Shanghai Municipal Bureau of Public Security,**
**Shanghai Municipal Finance Bureau,**
**Shanghai Municipal Commission of Construction and**
**Administration, Shanghai Municipal Greening Administration**
**and Shanghai Municipal Tourism Administration**

## 组长 | LEADER

### 郑时龄 | Zheng Shiling

中国科学院院士

Academician of Chinese Academy of Sciences

### 吴为山 | Wu Weishan

中国美术馆馆长

Director of National Art Museum of China

## 成员 | MEMBERS

**按照姓氏笔画排序**
**Sorted based on surname strokes**

### 马清运 | Ma Qingyun

美国南加州大学建筑学院院长

Dean of the USC School of Architecture

### 毛佳樑 | Mao Jialiang

上海市规划协会会长

Chairman of Shanghai Municipal City Planning Administration

### 支文军 | Zhi Wenjun

原同济大学出版社社长

Former President of Tongji University Press

### 王澍 | Wang Shu

中国美术学院建筑学院院长

Dean of the School of Architectural Art, China Academy of Art

### 王建国 | Wang Jianguo

东南大学建筑学院院长、城市规划研究院院长

Dean of School of Architecture, Southeast University and Urban Planning and Design School, Southeast University

### 王才强 | Heng Chye Kiang

**(新加坡 | Singapore)**

新加坡国立大学教授，环境与设计学院院长

Professor of National University of Singapore and Dean of School of Design & Environment NUS

### 北川弗兰 | Fram Kitagawa

**(日本 | Japan)**

国际策展人、大地艺术祭发起人

International curator and Initiator of Echigo-Tsumari Art Triennial

### 伍江 | Wu Jiang

同济大学副校长

Deputy President of Tongji University

### 朱子瑜 | Zhu Ziyu

中国城市规划设计研究院副总规划师

Deputy Chief Planner of China Academy of Urban Planning & Design

### 张永和 | Yung Ho Chang

国家“千人计划”专家，同济大学教授

Member of the Recruitment Program of Global Experts; Professor of Tongji University

### 张杰 | Zhang Jie

清华大学建筑学院副主任

Deputy Director of School of Architecture, Tsinghua University

### 李磊 | Li Lei

中华艺术宫副馆长

Deputy Director of China Art Museum

## 李向阳 | Li Xiangyang

上海视觉艺术学院美术学院院长

Dean of School of Fine Art, Shanghai Institute of Visual Art

## 李振宇 | Li Zhenyu

同济大学建筑与城市规划学院院长

Dean of College of Architecture & Urban Planning, Tongji University

## 李晓峰 | Li Xiaofeng

上海大学艺术研究院副院长

Deputy Dean of Art Institution, Shanghai University

## 杨劲松 | Yang Jingsong

中国美术学院美术馆执行馆长

Executive Director of Museum of Contemporary Art of CAA

## 杨奇瑞 | Yang Qirui

中国美术学院公共艺术学院院长

Dean of the College of Public Art, China Academy of Art

## 杨剑平 | Yang Jianping

上海大学美术学院副院长

Deputy Dean of Academy of Arts, Shanghai University

## 汪大伟 | Wang Dawei

上海大学美术学院院长

Dean of Academy of Arts, Shanghai University

## 沈迪 | Shen Di

上海现代建筑设计集团副总经理兼总建筑师

Deputy General Manager and Chief Architect of Shanghai Xian Dai Architectural Design (Group) Co., Ltd.

## 玛莎 · 索恩 | Martha Thorne

**( 西班牙 | Spain )**

普利兹克建筑奖执行理事

Executive Director of Pritzker Prize

## 芭芭拉 · 菲舍尔 | Barbara Fischer

**( 加拿大 | Canada )**

多伦多大学美术馆馆长

Director of University of Toronto Art Centre

## 尚辉 | Shang Hui

全国城市雕塑建设指导委员会艺委会副主任

Deputy Director of Art Commission, Urban Sculpture Development Instruction Committee of China

## 郑佳矢 | Zheng Jiashi

资深城市雕塑建设管理专家

Expert of Urban Sculpture Development and Management

## 郑培光 | Zheng Peiguang

上海城市雕塑艺术中心副主任

Deputy Director of Shanghai Sculpture Space

### 施大畏 | Shi Dawei

上海文联主席、中华艺术宫馆长

Chairman of Shanghai Federation of Literary and Art Circles and Director of China Art Museum

### 赵宝静 | Zhao Baojing

上海市城市规划研究院副院长

Deputy President of Shanghai Urban Planning & Design Institute

### 徐冰 | Xu Bing

中央美术学院教授

Professor of China Central Academy of Fine Arts

### 殷小烽 | Yin Xiaofeng

全国城市雕塑建设指导委员会艺委会副主任

Deputy Director of Art Commission, Urban Sculpture Development Instruction Committee of China

### 莫森·莫斯塔法维 | Mohsen Mostafavi

**( 美国 | USA )**

哈佛大学设计学院院长

Dean of the Harvard Graduate School of Design

### 曹嘉明 | Cao Jiaming

上海市建筑学会理事长

President of the Architectural Society of Shanghai

### 曼纽尔·库德拉 | Manuel Kudla

**( 德国 | Germany )**

国际建筑评论委员会秘书长，卡塞尔大学教授

General Secretary of International Committee of Architectural Critics ; Professor of University of Kassel

### 隋建国 | Sui Jianguo

国际著名艺术家

Artist

### 曾成钢 | Zeng Chenggang

中国雕塑学会会长

President of China Sculpture Institute

## 总策展人
## CHIEF CURATORS

### 伍江 | Wu Jiang

教授、博士生导师
同济大学副校长
法国建筑科学院院士
Professor, Doctoral Supervisor
Vice-president of Tongji University
Fellow of French Architectural Academy of Science

### 莫森·莫斯塔法维 | Mohsen Mostafavi

建筑师、教育家
哈佛大学设计学院院长
Architect and Educator
Dean, GSD, Harvard University

## 城市－建筑策展人
## CURATOR OF URBANISM-ARCHITECTURE

### 李翔宁 | Li Xiangning

同济大学建筑与城市规划学院教授、博士生导师，副院长
国际建筑评论家委员会委员、密斯·凡·德·罗奖评委
Professor and Deputy Dean, College of Architecture and Urban Planning, Tongji University
Member of International Committee of Architectural Critics (CICA)
Jury member of Mies Van der Rohe Award

## 艺术策展人
## CURATOR OF ART

### 张晴 | Zhang Qing

中国美术馆研究与策划部主任
同济大学、中国美术学院、中央美术学院、上海戏剧学院、云南大学兼职教授
Head of the Curatorial and Research Department of the National Art Museum of China
Guest Professor at Tongji University, China Academy of Art, Shanghai Theatre Academy, and Yunnan University

## 上海特展策展人
## CURATORS OF SHANGHAI SPECIAL EXHIBITION

### 章明 | Zhang Ming

同济大学建筑与城市规划学院建筑系副主任、教授
同济大学建筑设计研究院（集团）有限公司原作设计工作室主持建筑师
上海市建筑学会建筑创作学术部主任
Deputy director / Professor of Department of Architecture / CAUP/ Tongji University
Principal Architect of Original Design Studio / TJAD
Academic Director of Architectural Design of the Architectural Society of Shanghai China (ASSC)

### 张姿 | Zhang Zi

同济大学建筑设计研究院（集团）有限公司原作设计工作室设计总监
高级建筑师
Design director of Original design studio / TJAD
Senior Architect

### 王林 | Wang Lin

上海交通大学建筑系教授
城市规划博士
哈佛大学研究学者
Professor, Department of Architecture,Shanghai Jiao Tong University
PhD in Urban Planning
Harvard University Visiting Scholar

### 奚文沁 | Xi Wenqin

上海市城市规划设计研究院规划二所所长
历史文化和公共空间艺术中心主任
高级工程师
Director of Planning Dept.2 in Shanghai Urban Planning and Design Research Institute
Director of Historic Culture and Public Space Art
Senior Engineer

**助理策展人**
**ASSISTANT CURATORS**

--------------------

## 李丹锋 | Li Danfeng

同济大学建筑与城市规划学院博士候选人
冶是建筑工作室主持建筑师
PhD Candidate, CAUP, Tongji University
Principal architect, YeArch Studio

## 杨扬 | Yang Yang

美国加州大学洛杉矶分校建筑学博士候选人、助教
PhD Candidate and Teaching Associate in Architecture, University of California, Los Angels

## 田唯佳 | Tian Weijia

博士，同济大学建筑与城市规划学院助理教授
PhD, Assistant Professor of CAUP, Tongji University

## 江嘉玮 | Jiang Jiawei

同济大学建筑与城市规划学院博士候选人
PhD Candidate, CAUP, Tongji University

## 李凌燕 | Li Lingyan

同济大学建筑与城市规划学院博士候选人
《时代建筑》编辑部兼职编辑
同济大学跨媒体艺术与传播研究中心主任助理
PhD Candidate, CAUP, Tongji University
Part-time Editor, *Time+Architecture* Magazine
Assistant to the Director, Tongji University Cross-media Research Centre for Arts and Communication

## 吴彦 | Wu Yan

多伦多大学建筑景观与设计学院视觉系策展专业硕士，现为独立策展人
Master of Visual Studies Curatorial Studies, Daniels Faculty of Architecture, Landscape and Design, University of Toronto. Independent Curator.

## 许伟舜 | Xu Weishun

弗吉尼亚大学高等荣誉学士
哈佛大学设计学院建筑学硕士
Master of Architecture degree, Harvard Graduate School of Design
Bachelor of Science with Higher Honor , University of Virginia

## 李妍慧 | Li Yanhui

同济大学建筑设计研究院（集团）有限公司原作设计工作室，媒介主管
Media Director of Original design studio / TJAD

## 王明颖 | Wang Mingying

上海市城市规划设计研究院规划二所，高级规划师，项目总监
Senior planner and Project director, Planning Dept. 2, Shanghai Urban Planning and Design Research Institute .

## 高小宇 | Gao Xiaoyu

同济大学建筑与城市规划学院，博士候选人
PhD Candidate, College of Architecture and Urban Planning, Tongji University

## 徐继荣 | Xu Jirong

上海市城市规划设计研究院规划二所，项目负责人
Project director, Planning Dept. 2, Shanghai Urban Planning and Design Research Institute

CUSTOMIZED APPROACHES
TECHNIQUES ABSORBED

# 艺术人文视角下的
# 公共空间
# 与历史文化背景下的
# 城市更新

## 2015 上海城市空间艺术季策展感言

**伍江**
**同济大学副校长**
**2015 上海城市空间艺术季总策展人**

### I 背景

在经过近 30 年以大规模建设为特征的持续快速的城市扩张型发展之后，当下中国城市发展正处于重大历史转型期。以往大拆大建式的发展模式必将被新的发展模式所取代。城市建设以“量”的增长为主体正在转向以“质”的提升为主体。城市功能的进一步完善和优化、城市人文魅力的不断彰显和提升必将成为城市新一轮发展的主题。在这一转型过程中，城市空间的环境品质乃至艺术品质的塑造就必然成为新一轮发展的热点。近年来，北京、上海、广州、深圳、成都等国内各大城市，纷纷大张旗鼓地举办以城市建筑为主题的大型艺术展览活动，以艺术展的方式推动专业界、艺术界对于城市建筑领域发展模式的探索。上海近年来举办的各种各样的公共艺术活动，比如上海双年展、艺博会、西岸双年展等活动，也常聚焦城市、建筑和设计主题。特别是西岸双年展，围绕建筑和设计，邀请中外建筑界名师参展，在国内外产生了较大的影响。

2014 年，上海市规划与国土资源管理局为进一步综合已有的各类展览活动的优势，扩大规划建筑专业界和艺术界对于城市发展热点的参与，决定举办一次更高水平、更大规模、更加综合、更具特色的大型展示活动，并取名为“上海城市空间艺术季”，以期形成一个连续（双年）举办的、以城市空间为主题的、具有国内外重大影响的国际艺术展品牌。城市空间，尤其是城市公共空间的人文意义是城市空间存在的核心内涵。围绕城市空间的公共艺术展示活动，无疑可以极大地推动全社会对于这一重大话题的参与，并极大提高社会各界对于这一问题的认识水平。完全聚焦城市空间主题的大型艺术展示活动在国际上也是第一次。因此这样的活动对于规划界和建筑界，对于有兴趣探讨城市空间问题的艺术家，都具有极强的吸引力和号召力。

在上海市规划与国土资源管理局与上海市文化广播影视管理局携手领导下，经过一年多的精心策划和组织，以"城市更新"为主题的"2015 上海城市空间艺术季"终于拉开了帷幕。

## II 主题

作为第一届空间艺术季，将当今中国城市转型发展时期最热门也最重要的"城市更新"作为主题，使艺术季从一开始就成为紧紧抓住时代热点、具有鲜明时代特征的公共艺术活动。城市更新是城市的永恒主题。作为人类生活空间载体的城市，随着人类社会的发展，自然会处于连续不断的更新过程之中。正是这种持续不断的更新才使城市始终充满活力，有了生命。但这绝不意味着城市更新的过程是不断破旧立新的过程。文化发展更应是"护旧立新"的过程而非"破旧立新"的过程。作为人类文明物质载体的城市，人类文化的沉淀相当大一部分也正是通过城市来体现的。因此，城市更新最重要的内涵，是城市文化遗产的保护与城市文化的传承。其核心不是城市不断"翻新"，而是在保证城市已有文化积累得到充分保护与传承基础上的创新发展。当然这种以保护为特征的传承也绝不是福尔马林式的保护。那种要求一部分人放弃对于新生活的追求而成为另一部份人"历史回忆道具"的想法是不道德的。城市不可能也不应该凝固在某一历史时刻，城市里的每一个人都有权利追求美好生活。城市不断变化是必然的，不可阻拦的。但这种变化应是小规模渐变而非大规模突变（大拆大建）的，是充分关注人的记忆和情感需求的，是给予历史足够尊重的。这也正是城市更新之所以困难的原因。同时我们也确信，传统城市空间（包括建筑）经过适当改造是可以适应新的生活需要的。这也正是城市更新之所以有趣的原因。从这个意义上讲，城市更新的另一个重要内涵，是在尽可能保护延续原有城市空间的前提下，不断提升城市的空间品质。这个品质中既包含我们新创造的空间元素，也包含我们如何使空间的历史元素适应新的生活需求。城市是一个活的生命体，城市更新的目的不仅仅是希望我们的历史被保护好，我们更希望她仍然保持旺盛的生命力，而这生命力中，就包含着城市生活水平的不断提高，城市空间品质的不断提升。城市空间品质的提升，除了高品质的规划设计，也需要高品质公共艺术的介入。艺术介入可大大提高空间的品味，使得空间品质不仅体现在物质层面，更体现在精神层面。

## III 特点

这一次上海城市空间艺术季设计为一个主展和多个实践案例展组成。此外，还包括大量的论坛和报告会等公共活动，形成一个持续三个月的公共艺术活动。

主展由一条主线、四个板块和两个特展所组成。一条主线，就是对于 100 多年以来世界各国关于城市空间发展和城市更新的发展历史做了一次完整的演绎。这条演绎主线使本次展览具有了很强的学术性，使每一个参观展览的人都有机会全面了解到 100 多年以来城市更新发展的脉络、重要概念和典型案例。围绕着这条主线我们有向前向后两个板块，第一个板块"回溯"，是对历史的态度。参展作品中有各种历史保护的理念，展示历史保护的各种探讨。既有极少在国际艺术展示活动中露面的来自哈瓦那的案例，也有在我国目前城市历史保护和更新中少有触及的工人新村改造探索如曹杨新村案例。第二个板块"前瞻"，是对未来的探索。这里有关于生态都市主义的研究，也有上海 2040 规划的展望，更有城市"嗅觉地图"这样的"另类"。这两大板块，一个是关于我们怎么对待城市的昨天，一个是我们怎么对待城市的明天，串起来成为一条完整的脉络。第三个板块"影射"，从乡村发展的角度反视城市。分别由木、竹和钢建造的三座真房子在主展中颇为抢眼，通过三座为乡村而建的房子如何更加接近城市的本质，即生活的多样性、建造的真实性与文化的历史性。第四个板块“互动”，以艺术家的视角审视城市问题。其中艺术家马堡中干脆将一个原尺寸的巨大混凝土搅拌罐搬到了现场并使之不停转动，以表现停不住脚步的城市建设。而艺术家胡项城则以12个貌似放置精致展品然却空无一物的玻璃盒子，向观众隐喻了当代城市见物不见人的空虚。这四大板块，共同构成了关于城市更新主题的完整展示体系。全球背景的参展者及其作品，使得本次展览具有了很强的国际性与前沿性。

除上述四大板块外，主展中还特别安排了两个特展，一个是密斯·凡·德·罗奖25年回顾展，一个是上海城市更新案例特展。前者是第一次向国内观众全面展示密斯奖，可以使观众对欧洲近25年以来建筑设计成就及其价值取向有一个全面的了解。后者则对上海30年来城市保护与更新的发展历史作了一次全面的总结与展示。这也是第一次向公众全面展示上海在历史文化保护与更新方面的成就。上海在大规模快速城市建设中高度重视历史文化遗产的保护，是我国最早提出历史保护问题并最早建立了完整的保护管理体系的，这是非常了不起的。上海虽谈不上历史悠久，但城市的历史价值对于每一位市民而言，并不在于它有多悠久。有昨天就有了历史，就有了文化遗产。上海的实践值得学习和推广。

在主展之外，在上海的12个区县，同时还有15个现场的案例展。这既大大增强了本次艺术季的公众参与度，更使得艺术季真正成为一次遍布全市的盛会。

## Ⅳ反思

总览本次上海城市公共艺术季的全部作品，也有不少地方值得反思。在规划建筑专业界和艺术界多视角深层探讨城市更新问题的同时，显露出一些在参展作品中几乎未见涉及却又不应该被忽略的问题。第一个问题，是作品对于城市更新中社会弱势群体关注的集体缺失。毫无疑问，城市更新与发展，城市空间品质的提升，最终是以满足大多数人的需要、实现大多数人的利益为目标。然而，当我们关注“大多数”人的需要与利益的时候，那些“少数”而又处于社会底层的弱势人群的利益往往更容易被忽略。因此，对于城市空间乃至于城市功能的“公共性”考量往往同时也意味着对于弱势人群的忽视甚至牺牲。当我们为城市空间品质的不断提升而自豪的时候，我们是否关注到底层弱势人群所能分享的份额？在这里，城市空间的公平与公正就成了一个大问题。公平与公正，不仅在于社会全体对利益的共享与守卫，更在于少数弱势人群对于利益的分享甚至得到专门照顾，因为弱势人群比起其他人往往拥有更少的机会。因此，一个社会弱势人群难以分享到其好处的城市公共空间，甚或城市空间品质愈加“绅士化”而导致弱势人群越来越难以分享其好处的城市公共空间，其意义是要大打折扣的。对于这一问题思考的完全缺失，不能不说是这次艺术季参展作品的最大遗憾。

此外，参展作品对于当下城市更新现状的批判性不够，也是一大遗憾。这次艺术季的大部分参展作品对于现实的反思严重不足，即便是天然具有更强批判性的艺术家，此次参展作品也是多有轻松的戏谑而不见严肃的批判。还有，作为一次大规模的艺术活动，虽然在策划初期就定位在国际性、实践性与公众性，一个主展加10多个现场实践案例展（实为各具特色的公共艺术活动）也的确大大彰显了艺术季的公共性，然而就大部分参展作品而言，特别是主展馆内的参展作品而言，观众的互动与参与仍嫌不足。

作为第一次，2015 上海城市空间艺术季已达到甚至超过了预期目标。但这样的活动绝不应该只此一次。成功的开端使我们有信心期盼下一届艺术季的到来并获得更大的成功。上海城市空间艺术季应该也能够成为中国乃至世界围绕城市空间主题的最具吸引力、最具影响力、也最重要的公共艺术活动！

# Public Space from the Perspective of Arts & Humanities and Urban Regeneration in the Context of History and Culture

## Curatorial Statement for SUSAS 2015

**Wu Jiang**
**Deputy President of Tongji University**
**Chief Curator of SUSAS 2015**

### I Background

After nearly 30 years of steady and rapid development featuring large-scale construction and urban expansion, Chinese cities are currently undergoing a major historical transformation period. The "Mass-demolishing Mass-construction" model will definitely be replaced by a new way of development. Urban construction is shifted from quantitative increase to qualitative improvement. How to further improve the functions and services of a city and how to beautify the human urban environment will be the ultimate focus in the new chapter. The transformation process will emphasize on the building of environmental and artistic qualities of urban space. In the past few years, major cities in China including Beijing, Shanghai, Guangzhou, Shenzhen and Chengdu have organized large-scale art exhibitions on the theme of urban architecture to engage professionals as well as the art community in seeking and exploring new modes of development in urban architecture. Various public art events were held in Shanghai in recent years, including Shanghai Biennale, Shanghai Art Fair and West Bund Architecture and Contemporary Art Biennale, which often highlighted the themes of city, architecture and design. The West Bund Biennale of Architecture and Contemporary Art, in particular, has invited local and international renowned architects and designers to participate, and has achieved significant impact in the field.

In 2014, in order to consolidate the merits of existing exhibition events as well as to further engage the participation of professionals from the fields of architecture and art in the process of urban development, the Shanghai Municipal Bureau of Planning and Land Resources decided to organize a higher quality, larger scale and more comprehensive and unique exhibition event under the title of Shanghai Urban Space Art Season, with a vision of establishing an international brand of art exhibition with global influence. The humanistic significance is the core essence of urban space, particularly of urban public space. Public art events centered on urban space can effectively engage wide participation across the society and raise awareness of related issues. Internationally, this is the first art exhibition event of its kind that entirely devoted to the theme of urban space. Therefore, such an event has a strong appeal to the fields of planning and the architecture as well as to artists who are interested in the discussions around urban space.

Under the joint leadership of the Shanghai Municipal Bureau of Planning and Land Resources and Shanghai Municipal Administration of Culture, Radio, Film & TV, Shanghai Urban Space Art Season 2015 under the theme of Urban Regeneration is unveiled after over one year of meticulous planning and organization.

## II Theme

As the inaugural art season, this event is launched under the theme of Urban Regeneration, the most urgent and relevant topic during the current wave of urban transformation and development in China, making the art season a public art event that keeps pace with the times from the start. Urban regeneration is a lasting theme of urbanization. As the medium of human living space, cities are consistently in the process of urban regeneration in response to the evolution of human society. The on-going process generates and provides cities with endless energy and life. However, urban regeneration doesn't simply imply constant demolishing and construction. Rather, cultural development should be a process of preserving the old while creating the new as opposed to replacing the old with the new. As the material manifestation of human civilization, cities register the cultural sediment by humankind. Essentially, urban regeneration involves the preservation and inheritance of urban cultural heritage. What lies at the center is the adequate preservation of existing culture and innovative development predicated on inheritance rather than constant "refurbishment". Of course, such preservation oriented inheritance shouldn't be read as a formalin type of protection. It is an immoral practice to privilege the mere state of nostalgia while forbid the pursuit of a new life. A city should not be framed at a certain historical moment, every inhabitant is entitled to the right of pursuing a better life. The constant evolution of a city is inevitable. But the process of the evolution shall pace itself at a small-scale as opposed to large-scale mutation (mass-demolishing and mass-construction). It shall respect the history and appreciate the memory and emotional sensibility of humankind. This is the exact reason why urban regeneration is challenging. Meanwhile, with proper design and renovation, we believe that traditional urban space (including architecture) is able to adapt to new life styles. This is why urban regeneration is also interesting. That is to say, another important implication of urban regeneration is to improve urban spatial quality at the premise of maintaining existing urban spatial order. The quality consists of newly generated spatial elements as well as revised historical elements to adapt to the needs of new life styles. A city is a living entity. The objective of urban regeneration is not only to preserve our history, but also to stimulate her energy and vitality through the improvement of living standard and urban spatial quality. Apart from high-quality planning and design, the involvement of high quality public art is also required for the improvement of spatial quality in the city. In addition to sophisticated planning and good design, the intervention of high quality public art works can contribute to the improvement of urban space. Art can beautify urban space in both material and spiritual dimensions.

**III Characteristics**

Shanghai Urban Space Art Season 2015 consists of one main exhibition and various site projects. The three-month long art season also includes public events such as forums, discussions, and lectures.

The main exhibition is comprised of one key trajectory, four sections, and two special exhibitions. The key trajectory is a comprehensive interpretation of the historical development of urban space and urban generation around the world over the past 100 years. This line of interpretation involves intense academic researches and provides every visitor a great exposure to the context, important concepts and classic case studies around the development of urban regeneration over the past 100 years. Evolved from the key trajectory, two sections are developed to focus on the past and the future of our time respectively. First is "Retrospective", which reflects our attitude towards history. It introduces ideas and discussions around preservation of history, including case studies from Havana which are rare to be found in international art exhibitions and a case study of the transformation of workers' new village such as Caoyang New Village which hasn't been mentioned much in the current discussion of preservation and regeneration of urban history in China. The second section is "Prospective", which is to look into the future. It includes studies on ecological urbanism, outlook of Shanghai Master Plan 2040 and "alternatives" such as Smellscape Shanghai 2015. These two sections introduce how to position ourselves in relation to the past and the future of the city and delineate the context of it. The third section

"Reflection" tends to observe the city from the perspective of rural development. Three life-size houses made of wood, bamboo and steel respectively stand out in the main exhibition. Through introducing the houses originally designed for rural areas, this section tries to approach the essence of a city: diversity of life, authenticity of construction, and historical significance of culture. The fourth section "Interplay" focuses on urban issues from artists' perspective. Artist Ma Baozhong brings a full-size concrete agitation tank to the site and keeps it rotating to reflect the ceaseless process of urban development. Artist Hu Xiangcheng shows 12 empty glass display cases to suggest the missing objects which are supposed to be on display. It points to the reality of contemporary cities: empty objects with no human trace. Together, these four sections illustrate a coherent system of presentation to introduce the theme of urban regeneration. With participants from all over the world, this exhibition is indeed international and cutting-edge.

Apart from the four sections, two special exhibitions are arranged in the main exhibition. One is the retrospective exhibition "25 years of Mies van der Rohe Award" and the other is "Case Studies of Urban Space — Shanghai Urban Study". The former is a first comprehensive introduction of the Mies van der Rohe Award to Chinese audiences. It provides a rare exposure to the achievements of architecture design and the trends and shifts in the field of architecture over the past 25 years in Europe. The latter introduces the approaches of urban preservation and regeneration in Shanghai over the past 30 years. It is the first comprehensive public presentation of Shanghai's achievements in historical and cultural preservation and regeneration. Shanghai has invested great efforts in the protection of historical and cultural heritage in the process of large-scale and rapid urban construction. It is the first city in China that has raised the issue around historical protection and established a comprehensive protection management system. Shanghai doesn't have a long-standing history, but the value of history cannot be measured simply by its length. If there is a past, there will be historical and cultural heritage. The practices in Shanghai are worth to be studied and promoted.

Apart from the main exhibition, 15 site projects are spread across 12 districts and counties in Shanghai. It significantly improves the public engagement with the art season and transforms it into a city-wide celebration.

**IV Reflection**

Reflecting on all the works in the SUSAS 2015, there are things that need to be addressed. As an in-depth discussion on urban regeneration among the professionals from the fields of planning, architecture and art from multiple perspectives, some crucial issues which should have been addressed were absent in the exhibition. Firstly, there is a lack of attention to the socially underprivileged and marginalized communities which were left behind in the process of urban presentation. Undoubtedly, the ultimate goal of urban regeneration and the improvement of urban space quality is to serve the interests of the majority. However, when the focus is on the majority, the needs of those underprivileged ones stuck at the bottom of the society are most likely to be ignored. Therefore, the measure of "publicness" in urban spaces and civil functions doesn't take into account the interests of the socially underprivileged and marginalized communities. When we are proud of the improvement achieved in the urban space quality, do we ever concern how much of it is shared by the socially underprivileged and marginalized communities? Fairness and integrity of urban space becomes a big problem. Fairness and integrity is not only about balancing the interests across the society, but also about whether the socially underprivileged and marginalized communities is entitled to the equal share or not. The socially underprivileged and marginalized communities tend to have less access to opportunities. Therefore, there are less public spaces that are designed to attend their needs and the gentrification of urban spacc further distances the socially underprivileged and marginalized communities from accessing the benefits of urban public space. It is a shame that discussions around these issues are completely absent from the exhibiting works in this art season. Secondly, there is an apparent lack of critical voices reflecting on the current state of urban regeneration. Even the works by artists, who are supposed to be critical in nature, are more light-hearted jokes than serious criticism. In addition, as a large-scale art event, it is originally positioned to be an international, practical and public event. The approach of one main exhibition alongside a dozen of site projects (public art events in nature) does address the publicness of the art season, but most of the exhibiting works, especially the ones in the main exhibition, could have been more interactive and participatory with the audiences.

As the inaugural edition, Shanghai Urban Space Art Season has achieved or even gone beyond the original expectation. But it shouldn't end here. A successful start grants us the confidence to expect a second edition and aim for greater success. Shanghai Urban Space Art Season can and must become the most attractive, influential and important public art event centered on the theme of urban space in China and even across the world!

# 生态都市主义视角下的城市更新

**莫森·莫斯塔法维**
**哈佛大学设计学院院长**
**2015 上海城市空间艺术季总策展人**

城市化是当代社会面临的最重要问题之一。一方面，城市化是国家进步的标志，为改善人民生活提供无限可能。良好的住宅建设、高效的基础设施和交通系统、充足的工作机会、完善的公共空间和公共机构（比如公园、医院和大学）等，都是保证城市生活质量的基本元素。但另一方面，城市化进程和城市的运行机制却让它们成为地球上最大的能源消费者。在城市产生物质和精神文明的同时，它们也一并产生着垃圾和污染。

那么，我们如何视城市更新为契机和挑战，在改善城市生活体验的同时，不忘对市民权益和生态环境心存敬畏？

本次展览的目的之一就是用生态都市主义的角度看待这种挑战，将生态意识融入城市设计的创造力之中。

因此，我们需要用创新型的方法对待现有城市和新兴城市的开发，这些创新型方法必须在宏观和微观层面上行之有效，从城市图景到建造细节都不能遗漏。

从分流系统到食物生产，从公共空间到林荫便道，我们的城市是从物质、感官和艺术层面上提供市民幸福感的场所。我们要珍视所处的城市环境，并在它们的再造过程中尽自己的一份力。

# Urban Regeneration from an Eco-urbanism Perspective

**Mohsen Mostafavi**
**Dean of GSD, Harvard University**
**Chief Curator of SUSAS 2015**

Urbanization is one of the most crucial issues facing society today.

On the one hand, urbanization is a symbol of the progress of a nation, increasing opportunities for improving the lives of citizens. Good housing, efficient infrastructure and transportation systems, employment, public spaces and public institutions—such as parks, schools, hospitals and universities—are all examples of elements that help enhance the quality of our cities.

On the other hand, the processes of urbanization and the workings of our cities are amongst the highest users of energy and resources on the planet. Of course cities also produce things, both material and intellectual. But they also produce enormous amount of waste and pollution.

How can we take on the challenge of urban regeneration in such a way that can enhance our experience of urban living while also remaining respectful to the planet as well as the rights of citizens?

One of the aims of this exhibition is to respond to this challenge through the lens of ecological urbanism, an approach that brings together ecological awareness with creative urban design.

In this sense we need to reconsider the development of our new as well as existing cities through alternative and innovative approaches—approaches that work at both macro and micro levels—from the big picture to the small detail.

From distribution systems to food, from public spaces to the promenade, our cities are sites of physical, sensory and artistic pleasures. We need to cherish them and participate in their re-making.

SHANGHAI URBAN
SPACE ART SEASON
上海城市空间
艺 术 季

展场分区图 | The Exhibition Zoning Map

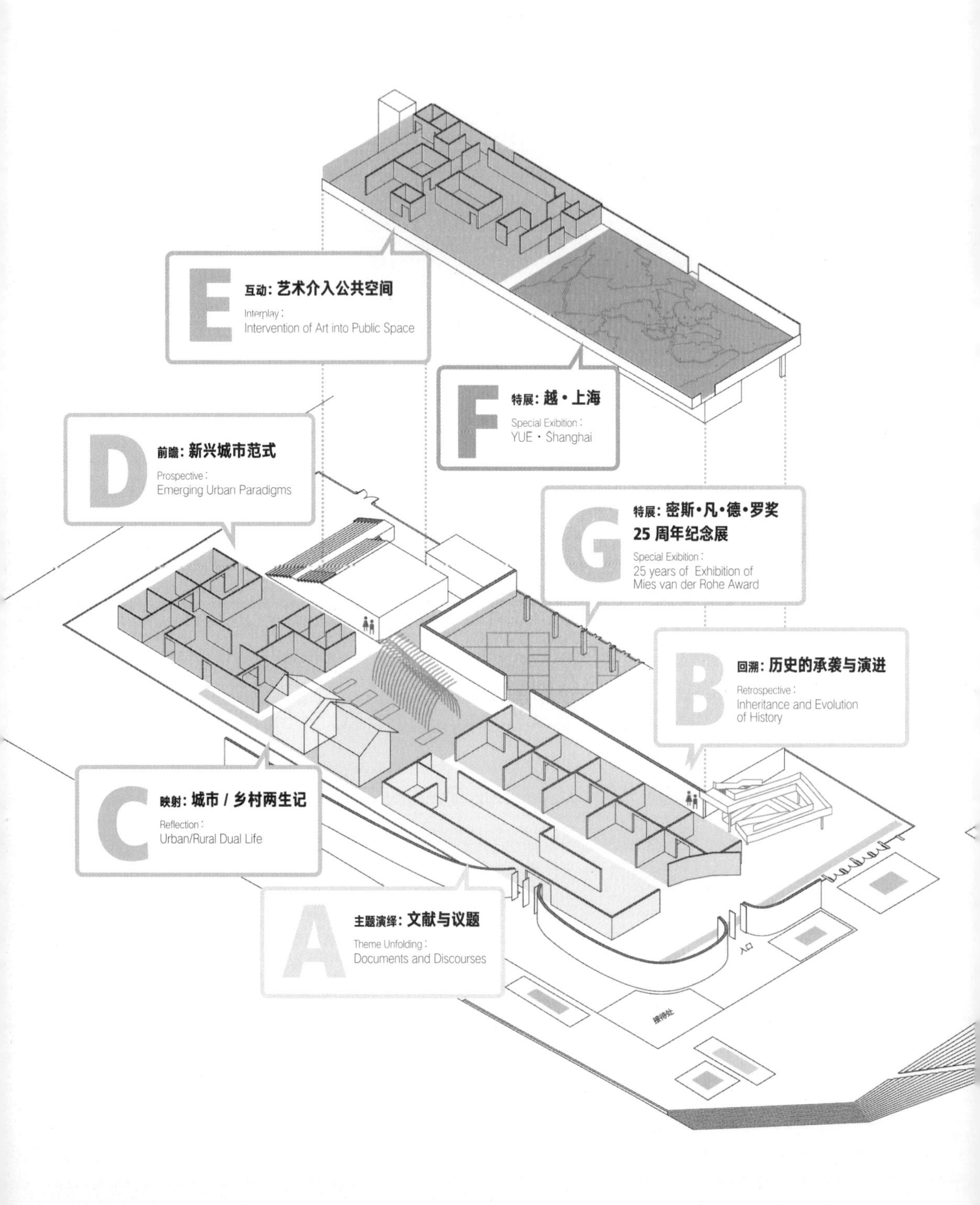
E
互动：艺术介入公共空间
Interplay：
Intervention of Art into Public Space
F
特展：越・上海
Special Exibition：
YUE・Shanghai
D
前瞻：新兴城市范式
Prospective：
Emerging Urban Paradigms
G
特展：密斯・凡・德・罗奖
25 周年纪念展
Special Exibition：
25 years of Exhibition of
Mies van der Rohe Award
B
回溯：历史的承袭与演进
Retrospective：
Inheritance and Evolution
of History
C
映射：城市 / 乡村两生记
Reflection：
Urban/Rural Dual Life
A
主题演绎：文献与议题
Theme Unfolding：
Documents and Discourses
入口
接待处

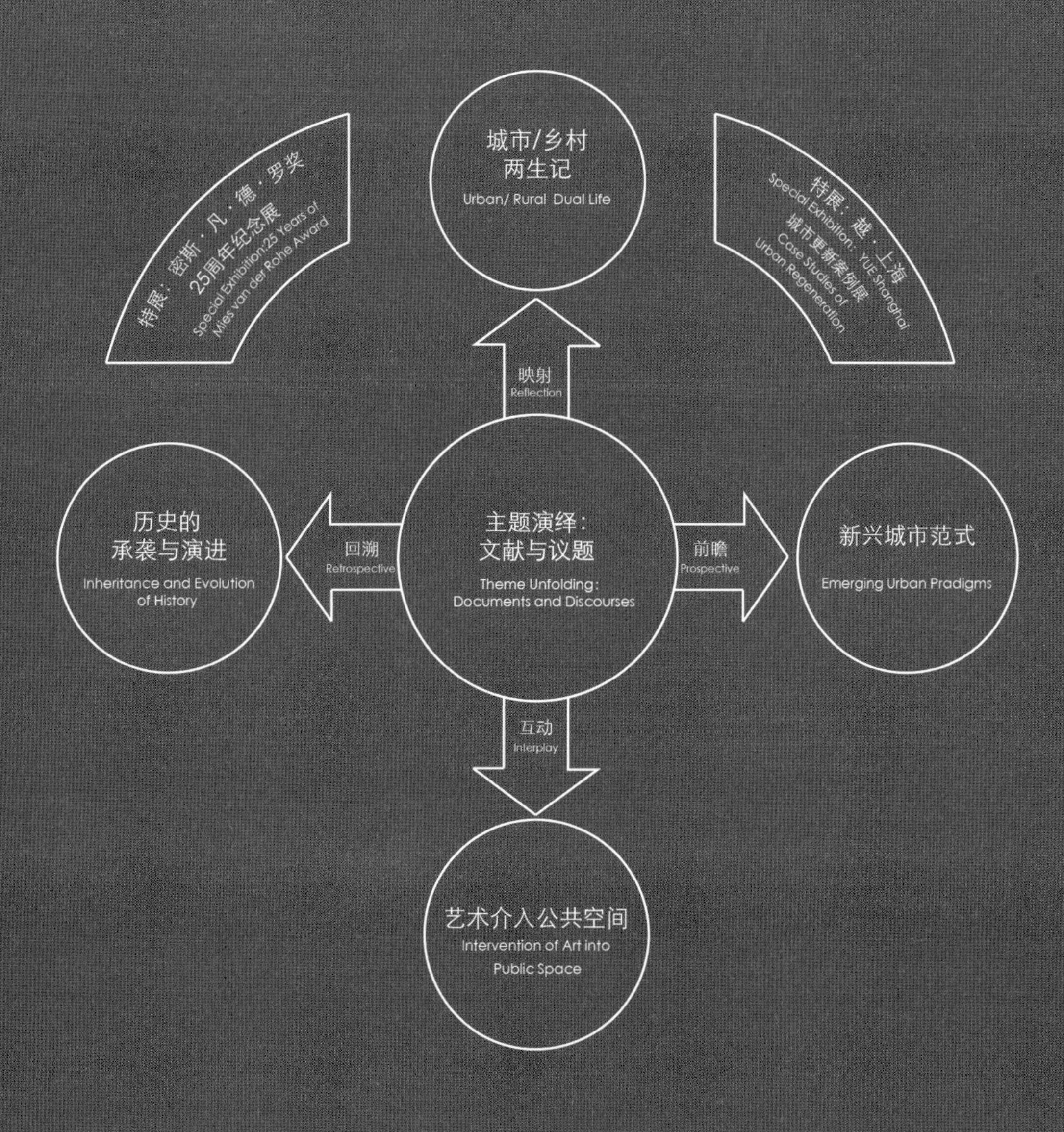
特展：密斯·凡·德·罗奖
25周年纪念展
Special Exhibition:25 Years of
Mies van der Rohe Award
城市/乡村
两生记
Urban/ Rural Dual Life
特展：越·上海
Special Exhibition：YUE Shanghai
城市更新案例展
Case Studies of
Urban Regeneration
映射
Reflection
历史的
承袭与演进
Inheritance and Evolution
of History
回溯
Retrospective
主题演绎：
文献与议题
Theme Unfolding：
Documents and Discourses
前瞻
Prospective
新兴城市范式
Emerging Urban Pradigms
互动
Interplay
艺术介入公共空间
Intervention of Art into
Public Space

# 主题演绎

## 文献与议题

# THEME UNFOLDING

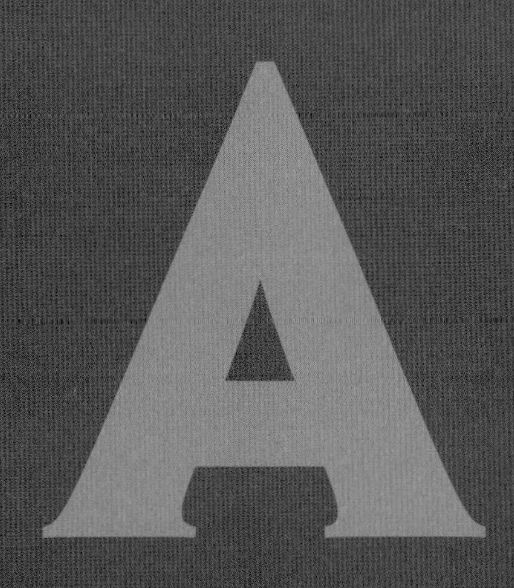

**研究团队**

李翔宁、杨扬、江嘉玮、谭峥、刘刊、丁凡、于云龙、张丹、吕凝珏、张向琳、韩叙、琚安琪、药乃奇、高长军、李野墨、吴越、陈迪佳、赵卓然、夏孔深、倪禛、邓圆也、伍雨禾、于炯主题演绎板块旨在以活泼、直观和互动的方式，向市民呈现关于“城市更新”的理论与实践的诸多视角。该板块将展示：学术文献、新闻报道以及法律法规等对于城市更新定义的概念演进；由城市管理者、规划师、建筑师以及最广大的普通市民对城市更新的理解与建议的访谈讨论；与城市更新相关的理论话语与议题的阐释，包括对机构、个人与所要解决的城市问题的介绍；关涉城市更新的重要立法、宣言、出版物、会议和危机事件等历史知识；城市更新经典案例的开发背景、战略目标、空间设计和管治模式等介绍。

## THEME UNFOLDING: DOCUMENTS AND DISCOURSES

"Theme Unfolding" aims at showcasing multiple perspectives of urban regeneration theory and practice through a vivid and interactive display. The section includes the evolving definitions of urban regeneration from scholarly publications, news and reports, as well as laws and regulations; interviews with city administrators, planners, architects, and the general public, about their thoughts and suggestions for urban regeneration; interpretations of relevant discourses and agendas, including involved institutions, individuals, and the targeted urban problems; historical knowledge of significant laws, manifestos, publications, conferences, and even crises; and an introduction of the developing background, strategic goals, spatial designs, and management models of typical cases of urban regeneration.

## 城市更新：内向的前线

李翔宁

当代城市化的演进，在现代主义的芝加哥模式和后现代的洛杉矶模式之后，中国向世界贡献了快速发展的新模式，这是一种和以洛杉矶为代表的都市蔓延式发展（Sprawl）相似但更高密度、更高速度的版本。如果说 20 世纪 80 年代以来近三个十年的中国城市狂飙式高歌猛进的时代需要一个减速的时段，那么我们已经渐渐看到这个时段的到来。

当城市摊大饼式拓张达到极限，我们不再有更多可以新建的空白用地，城市发展的前沿像水波纹的涟漪在彼此边界碰撞，但和水波彼此穿越的波纹不同，此时的城市只能转而向内，将前沿指向身体的内部。这种转向正是标志着城市更新的纪元开始。

正如我们为 2015 上海城市空间艺术季主展馆的策展前言中所述："城市，不仅是一个在空间上具有多种形式可能的开放领域，其政治经济与社会文化涵义在全球化过程中也愈加复杂和多变。城市更新，是针对活力衰退或失效的城市区域进行调适和再活化，从而解决城市问题、促进地区经济环境的可持续发展与社会品质的提升。"

艺术季主展览试图呈现的正是这样的视野，把当代世界城市更新的观念演变、话语和实践的经验呈现出来。展览的结构以主题演绎为核心，并通过两条轴线展开：一是回溯／前瞻的轴线，一是映射／互动的轴线，前者的两端分别是当代城市更新的既有案例和未来模式的展望，后者则通过城市／乡村的关系、艺术和建成环境的关系来丰富城市更新的内涵。与此同时，两个特展为整个艺术季展览增色不少，一是密斯·凡·德·罗欧盟当代建筑奖 25 周年回顾和 2015 年的评审结果特展；二是关注上海城市更新案例的"越／上海"。

### 主题演绎：文献与议题

"主题演绎：文献与议题"板块彰显了本届上海城市空间艺术季主展览的学术性与教育性，旨在以活泼、直观和互动的方式向市民呈现关于"城市更新"的历史、理论与实践的多重视角。该板块内容包括五大议题：概念、法案、话语、案例及访谈。

主入口的巨型雕塑集成了围绕城市更新不断演进发展的概念关键词，悬浮于抽象的城市雕塑上空，隐喻本届艺术季主题的综合性与多维度。板块共收录了学术文献、新闻报道、法律法规以及专家访谈中对城市更新的一百多个定义，对这一世界性议题的概念进行了历史的回顾与当代的展开。

板块展区主墙面分别梳理了两条历史脉络：1）关涉城市更新的重要立法、宣言、出版物、会议和危机事件等历史知识；2）与城市更新相关的理论话语与议题的阐释，包括对机构、个人与所要解决的城市问题的介绍。板块的主要展示区域以历史阶段、实践类型、城市资源、应对策略等类别分组展示了城市更新经典案例的开发背景、战略目标、空间设计和管治模式等信息。并通过三维空间的连线展示了法案、话语与案例之间深远的历史关联与影响，力图向观众全面地勾勒和呈现本届展览主题的丰富内涵与其之于我们生活城市的尚待发掘的潜质。

纪录影像“都市漫谈 /Urban Gossip/ 城里厢嘎讪胡”对上海不同片区，包括热门旅游景点、老城区、商业中心、动迁区等多种类型城市空间中的本地居民、学生、外来务工人员、游客等多种人群进行了街头访谈，以生动诙谐的形式记录了大众对上海这座城市的印象、体验、理解与寄语。而对 7 位包括策展人、规划师和建筑师在内的专家访谈则从各自领域就其对城市更新的实践与研究发表了观点与建议。“主题演绎”板块在面向公众的同时，也为当代的学者、实践者与城市管理者提供了促进交流与对话的平台。

**从回溯到前瞻：城市更新的前世与未来**

“回溯”板块旨在对城市历史保护与更新的相关实践与理论研究进行回溯，呈现地方、国家和全球语境下的新问题与城市策略。

该板块将展示中国古城古村落的建筑修复和设施更新的经验成果，包括北京华清安地建筑设计事务所有限公司主持的成都宽窄巷子保护设计、同济大学与威尼斯建筑大学共同参与的上海曹阳新村社区公共空间研究。同时还带来许多国内外历史街区保护与历史建筑改造和文化传承的优秀案例，比如米拉莱斯塔格里亚布事务所主持的汉堡港更新设计、同济大学与哈瓦那大学共同参与的哈瓦那旧城更新课题研究、西班牙克鲁斯奥蒂斯事务所负责的荷兰国立博物馆更新设计、挪威奥斯陆建筑学院名为“定制”的研究课题、洛加大建筑学院城市实验室的上海大都市更新问题研究、11 位北京设计师的 11 份北京城市更新作品、12 家上海建筑事务所的 12 件上海在地更新作品、同济大学负责的上海陆家嘴城市研究与“微缩之城”上海百年历史演变研究、巴塞罗那政府主持的巴塞罗那城市公共空间更新计划、西班牙加泰罗尼亚理工大学建筑学院的历史城市课题回溯等。

与此同时，我们的研究还试图呈现艺术运作、文化机构与城市互动为动力的城市复兴的关系。比如加州大学伯克利分校艺术 + 村落 + 城市研究工作室对珠江三角洲艺术家村落的研究作品精选、同济大学建筑与城市规划学院砖雕与木雕实践作业展等。共同展示对当下城市历史保护问题的多维度思考、应对快速城市化的多重策略、以及新的生活方式给旧城更新带来的发展契机。

“前瞻”板块将展示数字媒体文化对建筑与都市想像的催生与影响。媒体文化不仅记录和展现城市与建筑的实践状况，并且推动理论、评论和学科的发展，更直接参与社会生产方式的变革，对于建筑与都市具有显著的催化作用。“前瞻”板块旨在展示信息化时代下，新传媒、计算机参数化设计、互联网技术等对新的都市生活范式和城市公共空间的探索与引领。

这些研究有的从生态和能源角度出发，例如日本阵内秀信研究室（法政大学）等团队组成的“东京二零五零”和上海同济城市规划设计研究院的“上海二零四零”，两个团队不约而同地从如何有效利用生态与能量这个议题而展开探讨，对都市构造及生活形态的重塑提出方案。大数据是另一种未来城市发展所运用的重要基础，在达尼尔·戈德迈耶等互动式媒介研究专家带来的“百老汇”街的研究成果中，研究者沿着百老汇街中心每隔 30 米布点，并以该点为中心做一个 100 米宽的切片，结果就是一个 21 390 米长、100 米宽的脊椎形态，用这个形态坐标来过滤取自整座纽约城的数据。总体上看，这项研究使用了超过 4 千万张图片和数据点来再现了一条街道。此外还有上海西岸开发（集团）有限公司带来的“上海西岸传媒港”、哥伦比亚大学建筑、城规与保护研究生院带来的“能量试点工程”、米兰理工大学城市学博埃里教授带来的“垂直森林&森林之城”、ASA 建筑设计事务所带来的“天空花园”以及加州大学洛杉矶分校研究工作室带来的“色彩乌托邦：上海的多样高楼”、何塞·潘纳拉斯等建筑师团队带来的“感觉马德里”、都市实践建筑设计事务所带来的“城市更新中的都市实践”、哈佛大学设计博士范凌带来的“科技的 8 个设计议题”。这些团队从不同地域案例出发，多元角度地展示了数字化和媒体化对建筑与都市语法形态的深厚影响，给观者带来更为直观的体验。

**映射与互动：城市／乡村两生记与艺术的介入**

“映射”部分的主题是城市 / 乡村两生记。探讨城市与乡村互为依存、互为参照的复杂关系。当下中国的都市实践的视野不再囿于城市。在政策与公共事件背后，乡土实践越来越得到青年建筑师和学者的关注。

通过展示 2000 年以来一批优秀中外建筑师的乡土实践、研究学者对未来乡村发展的类型与原型研究、以及艺术家为延续乡村社区文化而尝试的社会性介入，“映射”板块旨在回应过度城市化与失衡的城乡关系，思考乡村哲学，折射对都市建设的反思，并探讨未来城乡融合互动实践在政治经济层面展开的可能性。城市与乡村的面孔在这次展览中不再各自独立，而是真正呈现出一种相互作用的状态。

为了更加直观的展示这种状态，本次展览将建造本身作为展示的内容。展场中央的三处搭建作品，分别来自于谢英俊、王灏和陈浩如三位建筑师。他们分别以钢、木、竹为主要材料，在展览开幕前现场建造了三个别具特色的结构体，将各自多年来的实践视角清晰地展现出来。在三个作品内，展示了渠岩、左靖等艺术家们的乡村实践成果。而在三个作品之间，则是一批优秀的乡土建筑作品，如华黎的武夷山竹筏育制场等。可以感受到的是，当城市价值观念进入乡村后，通过众多乡建工作者的努力推动，乡村的价值观念正在以一种强劲的势头被反馈给城市。二者在思想观念上的趋于平等，是解决城乡空间、资源分配、居民生计等问题的关键所在。是城市与乡村的相互读取和传输，才使得这些乡建作品可以活生生地进入博物馆空间，带着各自的经验和矛盾，等待人们挖掘新的价值和发展的道路。

而“互动”板块（艺术介入城市生活）则呈现了艺术家眼中的城市更新。由张晴策展这个板块，集中了 34 件国际、国内艺术家的作品。他们或以艺术的手段和视点呈现城市更新的状态，或者通过艺术的介入对城市更新的意涵进行令人深省的发问。失去了艺术的城市将是没有未来没有生机的城市，艺术正是城市更新的巨大动力。

**密斯·凡·德·罗奖 25 周年纪念展**

33 年前的 1982 年，借巴塞罗那复建密斯 1929 年世博会德国馆之契机，密斯·凡·德·罗基金会成立。五年后的 1987 年，巴塞罗那市政府和欧盟共同签署了一项奖项协议，用于对欧洲建筑文化进行表彰与推广，这就是如今全球最重要的建筑奖之一——欧洲建筑密斯·凡·德·罗奖的由来。

自 1988 年授予葡萄牙建筑师西扎的作品平托 & 索托银行起，密斯·凡·德·罗奖已经度过了 27 个春秋。不同于对建筑师职业生涯做综合性评定的普里兹克奖，密斯奖是一个”以作品论英雄“的奖项，在对项目进行参评时，所有评审需要同时听取来自设计师、所有者甚至使用者等多方意见；参展项目没有任何功能，大小等方面的限制，只要该项目由欧洲建筑师在欧洲境内设计，并在过去两年中建成的，皆有资格参评，历届获奖者中实不乏鲜为人知者。从这个角度上讲，密斯奖给了很多人一个可以打破“论资排辈”的机会。2015年是第14届的密斯奖，共有420个作品被提名。随后，由来自意大利的建筑师领衔的评审团从这420个项目中进行筛选。在确定5个最终获奖榜单后，诸位评审又亲临现场进行实际感知，并听取了设计者，甲方和使用者三方的评述。

名为“欧洲制造”的密斯奖特展，为我们带来了186个项目模型，3000余幅项目图纸，近30个视频和访谈。展览包括对于历年作品的回顾和2015年评选的详细介绍，信息量之大，在国内所有专业展览中也属罕见。在这里，一部简约的欧洲当代建筑史将向你和盘托出。

**城市更新：跨越界限的新前线**

作为跨学科的城市研究、政策与实践，城市更新涵盖政治经济文化的众多范畴，涉及可持续、社会公平、公共利益和效率等诸多议题。它既是全球城市治理面临的共同问题，亦提供了再思考城市概念本身的机会。

城市更新，已经成为不同学科、不同国家、地区和城市的共同关注。与2015上海城市空间艺术季几乎同时举办的深圳双年展也把“城市原点（RE-LIVING THE CITY）”作为主题，同样是探讨在城市发展的新阶段，回到城市的原初，重新审视既有的城市区域，重新找回城市发展的原动力，使得我们已有的城市家园变得更美好。

城市更新是一个新的纪元，让我们可以重新省思我们的建成环境，检视我们在城市建设和发展中曾经的轻率误判，给我们一个打破既有城市体系和格局，重塑公共空间、艺术形式乃至生活方式的机会。这是一个转向自身，再次启航的新的前线。

## Urban Regeneration: Development Frontier Turning Inward the City

**Li Xiangning**

With respect to the evolution of contemporary urbanization, China has contributed a new model of rapid development to the world after the modern Chicago Model and post-modern Los Angeles Model. This is a development version that is similar to the urban sprawl development represented by Los Angeles but with higher density and faster speed. If Chinese cities need a slow-down period after nearly three decades of rapid development since the 1980s, we have seen the gradual arrival of this period.

When unorderly urban expansion reaches its limit, more idle land for new buildings is no longer available. The forefronts of urban development are like water ripples colliding at the border. However, unlike water ripples run across each other, a city can only turn to inward development, pointing the development frontier to the internal city. This shift marks the beginning of an era of urban renewal.

As we stated in the curator's preface for the main exhibition of the Shanghai Urban Space Art Season 2015, "City is not only an open field with a variety of spatial forms, but more importantly with increasingly complex and changeable political, economic, social and cultural meanings in the globalization process. Urban renewal refers to the adjustment and reactivation of urban areas whose vitality have been declined or lost, so as to solve urban problems, promote sustainable development of regional economy and improvement of social quality".

The main exhibition of Art Season attempts to present this vision and the concept, discourse and practical experience of urban renewal in the contemporary world. The exhibition is centered on theme development and unfolded along two axes, i.e. Retrospective/Prospective axis and Mapping/Interaction axis. The former focuses on the existing cases and future models of contemporary urban renewal, and the latter enriches the connotation of urban renewal through the relationship between cities and villages and the relationship between art and built environment. Meanwhile, two special exhibitions have added a lot of highlights to the whole exhibition of the Art Season. One of them is the 25th anniversary exhibition for the European Union Prize for Contemporary Architecture: the Mies van der Rohe Award and assessment results in 2015; the other is "More/Shanghai" exhibition that focuses on urban renewal in Shanghai.

## Theme Unfolding: Documents and Discourses

"Theme Unfolding: Documents and Discourses" section highlights the academic and educational objectives of the main exhibition of the Shanghai Urban Space Art Season this year, and aims to present the history, theory and practice of urban renewal to citizens in a lively, intuitive and interactive manner. This section includes five topics, i.e. concept, law, discourse, case and interview.

The giant sculpture at the main entrance has integrated the concept keywords about the constant evolution of urban renewal and suspended over the abstract urban sculpture, implying the comprehensive and multi-dimensional theme of this Art Season. This section has collected more than one hundred definitions of urban renewal in academic literatures, news reports, laws & regulations and expert interviews. Historical review and contemporary interpretation have been conducted on this worldwide topic.

The main exhibition wall of this section has followed two historical clues: 1) historical knowledge regarding important legislations, declarations, publications, conferences and crisis events concerning urban renewal; 2) theoretical discourse and topic interpretation concerning urban renewal, including the introduction of institutions, individuals and urban problems to be solved. The main display area of this section shows the development background, strategic objectives, spatial design and governance model of classical cases of urban renewal in such categories as historical stage, type of practice, urban resources and coping strategies. The line connection in three-dimensional space shows the far-reaching historical connection and impact among legislation, discourse and case, trying to fully outline and present to the audience the rich connotation of the theme in this exhibition as well as untapped potential of our cities.

The documentary video "Urban Gossip" shows street interviews with local residents, students, migrant workers and tourists in various types of urban spaces including popular tourist attractions, old towns, business centers and resettlement areas in Shanghai, and vividly records the impression, experience, understanding and message of the public about the city . The interviews with 7 experts including curators, planners and architects reveal their opinions and recommendations from their respective fields on the practice and research of urban renewal. The "Theme Unfolding" section is open to the public and also serves as a platform to promote exchanges and dialogues among contemporary scholars, practitioners and city administrators.

## From Retrospective to Prospective: Past and Future of Urban Regeneration

Retrospective section is intended to review the relevant theory and practice of urban conservation and renewal, presenting new problems and urban strategies in the local, national and global context.

This section will showcase experiences and

achievements of building renovation and facilities renewal in China's ancient cities and villages, including Chengdu Kuanzhai Alley Conservation Design conducted by An-Design Architects Co., Ltd., and study on public space in Shanghai Caoyang New Countryside Community jointly undertaken by Tongji University and Università IUAV di Venezia. This section also brings many excellent cases of historic district protection, historic building renovation and cultural heritage at home and abroad, e.g. Hamburg Port renewal design undertaken by Miralles Tagliabue Architectural Firm, Havana old city renewal research jointed conducted by Tongji University and University of Havana, Netherlands National Museum renewal design conducted by Spain Cruzy Ortiz Firm, research topic titled "Custom made" of Architectural Institute of Oslo, Norway, Shanghai metropolis renewal research conducted by Urban Laboratory of the School of Architecture of UCLA, 11 Beijing urban renewal works created by 11 Beijing designers, 12 Shanghai urban renewal works created by 12 Shanghai architectural firms, Shanghai Lujiazui urban study and centuries-old historical evolution study of "miniature city" Shanghai undertaken by Tongji University, Barcelona urban public space renewal plan sponsored by Barcelona city government, and retrospective study of historical urban topics carried out by the School of Architecture of Polytechnic University of Catalonia, Spain, etc.

Meanwhile, this section also attempts to present the relationship among artistic operation, cultural institution and urban renaissance driven by interaction among cities. For example, the Art + Villages + Urban Research Studio of the University of California at Berkeley shows the fine works of artist village in the Pearl River Delta, and College of Architecture and Urban Planning of Tongji University shows the practical works of brick carving and wood carving. These works will collectively demonstrate multi-dimensional thinking on the current issue of urban conservation, multiple strategies to cope with rapid urbanization, and development opportunities brought by new lifestyle to the old city renewal.

The Prospective section showcases the impact of digital media on architecture and urban imagination. Media culture can not only record and display the practical situation of the city and buildings, promote the development of theory, comments and discipline, but also directly participate in the transformation of social production mode, and have a significant catalytic effect on buildings and cities. This section aims to showcase the exploration and guidance of new media, computer parametric design and Internet technology for new paradigm of urban life and urban public space in the information age.

Some of these studies are conducted from the perspectives of ecology and energy. For instance, "Tokyo 2050" team from Hidenobu JINNAI Lab. (Hosei University) and "Shanghai 2040" team from by Shanghai Tongji Urban Planning and Design Institute discuss the topic of how to effectively use the ecology and energy, and propose schemes for urban construction and lifestyle reshaping. Big data is another important foundation for future urban development. Among the "Broadway Street" research achievements made by interactive media research experts including Darnell Godemeyer, the researchers arrange points every 30m along the center of Broadway Street, make a slice of 100m wide centered on this point, and thus form a spine of 21,390m long and 100m wide. The coordinates of this form are then used to filter the data from the entire New York City. Overall, this study has used more than 40 million images and data points to reproduce a street. Other studies include "Shanghai West Bund Media Port" of Shanghai West Bund Development (Group) Co., Ltd, "Energy Pilot Project" of the Graduate School of Architecture, Urban Planning and Conservation of Columbia University, "Vertical Forest & Forest City" of Professor Boeri in urban studies from Milan Polytechnic University, "Sky Garden" of ASA Architectural Design Firm and "Color Utopia: Shanghai's Diverse High-rise" of the research studio of the University of California at Los Angeles, "Feeling Madrid" of architects including Jose Pannalars, "City Practice in Urban Regeneration" of Urbanus Architectural Design Firm and "8 Design Issues in Science and Technology" of Doctor of Design Fan Ling from Harvard University. These teams proceed from cases in different regions, reveal the profound impact of digital and media technology on architecture and urban morphology from multiple angles, and bring to the viewer a more intuitive experience.

**Mapping and Interaction:**
**City/Country Record and Involvement of Art**

The Mapping section is themed on city/country record, exploring the complex interdependent and mutually referential relationship between the city and country. The vision of urban practice in contemporary China is no longer confined to the city. Behind the policy and public events, rural practice has drawn more and more attention from young architects and scholars.

By showing the rural practices of outstanding domestic and foreign architects since 2000, researches on the types and prototypes of the future rural development as well as social involvement attempts made by artists for the cultural continuation of rural communities, the Mapping section aims to reveal excessive urbanization and imbalanced urban-rural relationship, ponder over rural philosophy, show reflection on urban construction and explore the possibility of urban and rural integration practice at the political and economic level. The city and countryside are not independent from each other in this exhibition, but really show a state of interaction.

In order to more intuitively display this state, this exhibition uses the building itself as the exhibition content. Three building works in the center of the exhibition venue are from three architects, Hsieh Ying-Chun, Wang Hao and Chen Haoru. They have respectively used steel, wood and bamboo as main materials to construct three distinctive structures before the opening of the exhibition, and clearly demonstrated their years of practice. These three works have shown the rural practice of such artists as Qu Yan and Zuo Jing. Among the three works, there is a batch of outstanding works of rural architecture, such as Huali's Wuyi Mountain Bamboo Raft Factory and so on. You can feel that when the values of the city enter the countryside, the values of the village are showing a strong momentum of feeding back to the city through the efforts of many rural constructers. The equal ideology of both values is the key to solve such problems as urban and rural space, resource allocation and residents' livelihood. The mutual communication and transmission between urban and rural areas makes it possible for these rural building works to enter the museum, carry their experiences and contradictions and wait for the public to tap new values and development paths.

The Interaction section (art involves in urban life) shows the urban renewal in the eyes of artists. This section is curated by Zhang Qing, and has gathered 34 works of international and domestic artists. They either present the status of urban renewal by artistic means and viewpoints, or question the implications of urban renewal after art involvement. A city without art will be a city without future or vitality. Art is an enormous power for urban renewal.

**25 Years of Mies van der Rohe Award**

In 1982, 33 years from now, the Mies van der Rohe Foundation was established amid the reconstruction of the German pavilion of World Expo 1929 designed by Mies in Barcelona. Five years later, in 1987, the Barcelona City Council and the EU signed an award agreement for recognition and promotion of architectural culture in Europe. This is the origin of one of the world's most important architectural award, the Mies van der Rohe Award.

The Mies van der Rohe Award has a history of 27 years since it was first awarded to Portuguese architect Siza in 1988 for his Pinto & Soto Bank. Unlike the Pritzker Prize that makes a comprehensive assessment of architects' career, Mies Award is only presented to those designers with best building works. During project assessment, all assessors should listen to the opinions from various parties, including designers, owners and even users. There are no restrictions to the function and size of the project to be assessed. Any project is eligible as long as the project is designed by European architects in Europe and completed in the past two years. Previous winners have seen some rarely-known works. From this perspective, Mies Award has created an opportunity for a lot of people to break the "seniority" system. 2015 witnesses the 14th session of the Mies Award, with a total of 420 works nominated. Subsequently, the Italian architect Cino Zucchi leads the jury to screen from 420 projects. After a list of five final entries was determined, all assessors went to the sites and listened to the comments from designers, owners and users.

The special exhibition for Mies van der Rohe Award titled "Made in Europe" brings us 186 project models, more than 3,000 pieces of project drawings, nearly 30 videos and interviews. The exhibition includes review of works in the previous years and detailed assessment process in 2015. The mass information of this exhibition is rarely seen in all professional exhibitions in China. Here, a concise history of European contemporary architecture will be presented to you.

Urban renewal is a new opportunity for us to reflect upon our built environment, review our misjudgment in urban construction and development, break through the existing urban system and structure, and reshape our public space, art forms and even our way of life. This is a new frontier to turn inward the city and set sail again.

**Urban Regeneration.**
**New Frontier across Boundaries**

As an interdisciplinary urban research, policy and practice, urban renewal covers numerous fields including politics, economy and culture, and involves a number of issues such as sustainable development, social justice, public interest and efficiency. It is not only a common problem in regard to urban governance around the world, but also provides opportunities to rethink the concept of city itself.

Urban renewal has become the common concern of different disciplines, different countries, regions and cities. Shenzhen Biennale held almost simultaneously with the Shanghai Urban Space Art Season 2015 also takes the "Re-living the City" as the theme, explore the ways to return to the origin of the city, re-examine the existing urban areas and rediscover the driving force of urban development in a new stage of urban development, so as to make our city a better place to live.

Environment Oriented
Networking Society
Cultural Industry
City Marketing
Technocity
城市营销
新步行主义
沟通规划
Creative City
New Pedestrianism
土地再整理
Placemaking
城市复兴
Communicative Planning
Adaptive Reuse
包容性区划
景观都市主义
技术城市
Adaptive Reuse
Compact City
Urban Catalyst
都市农业
Creative City
事件导向
保护性再利用
文化产业
Urban Agriculture
Landscape Urbanism
社区行动主义
New Urbanism
环境导向
城市更新
精明增长
Smart Growth
Urban Renaissance
城市营销
文化导向
沟通规划

# A-01 城市更新概念演进 EVOLUTION OF THE CONCEPT OF URBAN REGENERATION

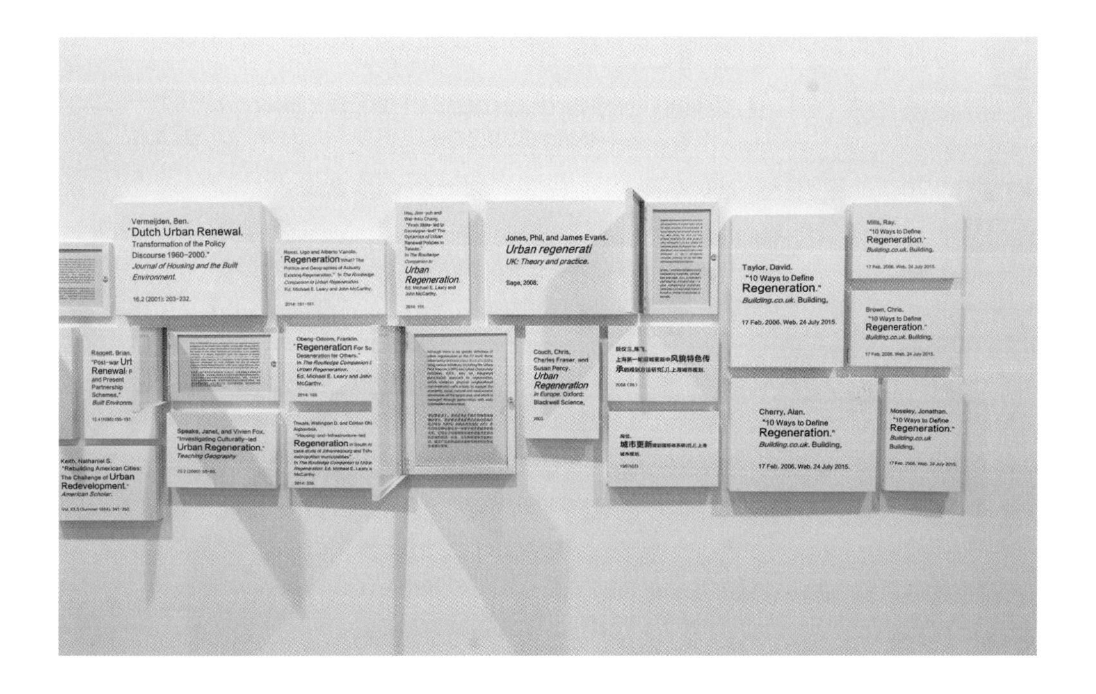


本参展作品回顾城市更新理论在西方城市史背景下的出现意义，并梳理城市更新理论话语的历史演变，从而追逐该议题对当今城市发展的作用。梳理学术文献、新闻报道，以及法律法规等对于城市更新定义的概念演进，以活泼、直观和互动的方式向市民呈现关于“城市更新”概念理解的诸多视角。

**This section reviews the emergence and evolution of urban regeneration theories within the context of Western urban history, and traces the significant impact it has on contemporary urban development. Through the presentation of academic publications, news reports, laws and regulations, this section provides dynamic perspectives and interpretations on the evolution of the concept of urban regeneration in an expressive and interactive way.**

社会资本
Social Capital

场所的空间
Space of Place

y
vitalization/
habilitation
复苏 内城振兴

Network Society
网络社会

Space of Flow
流动的空间

Edge City
边缘城市

Zwischensta
城市间

lic Housing
公共住房

nce
民窟

dder
阶梯

福利社会
Welfare State

保障性住房
Affordable Housing

士绅化
Gentrification

Urban Redevelopment
城市再开发

邻避症候群
NIMBY

Brownfield Renewal
棕地更新

Exopolis

Social Reconstruction
社会重构

沟通规划理论
Communicative Planning Theory

Inclusionary Zoning
包容性区划

Cultural Industry
文化产业

Public Participation
公众参与

ommunity Activism
社区行动主义

Cultural Cluster
创意集群

社区艺术运动
Community Arts Movement

Advocacy Planning
倡导性规划

威尼斯宪章
Venice Charter

Cultural Planning
文化规划

Civil Society
市民社会

City Marketing
城市营销

Urban Catalyst
城市触媒

Creative
创意

e
vation
保护

Authenticity
原真性

Placemaking
场所营造

Theme-parking
主题公园化

毕尔巴鄂效应
Bilbao Effect

Generic City
广谱城市

Megastructure
巨构

Junkspace
垃圾空间

完整性
Integrity

Plug-in City
插入式城市

拥挤文化
Culture of Congestion

Pragmatism
实用主义

Theory of B
“大尺度”理

Growth Machines
增长机器

Urban Entrepreneurialism
都市企业主义

Central Place Theory
中心地理论

Area-based Initiatives (ABIs)
地方性举措

Land Readjustment
土地再整理

Ente
企

功能性城市
Functional City

Intensive Land use
土地集约利用

Compact city
紧凑型城市

滨水区域复兴
Waterfront Revitalization

Adaptive Reuse
保护性再利用

棕地再开发
Brownfield Redevelopment

TND
传统住区开发

La Synthese Ecologique
生态学概论

New Urbanism
新城市主义

Urban Agriculture
都市农业

TOD
公共交通导

Design with Nature
设计结合自然

Sustainable City
可持续城市

绿色建筑
Green Building

Landscape Ecology
景观生态学

CNU

Urban Regeneration
城市再生

# A-02 城市更新话语地图 DISCOURSE MAP OF URBAN REGENERATION

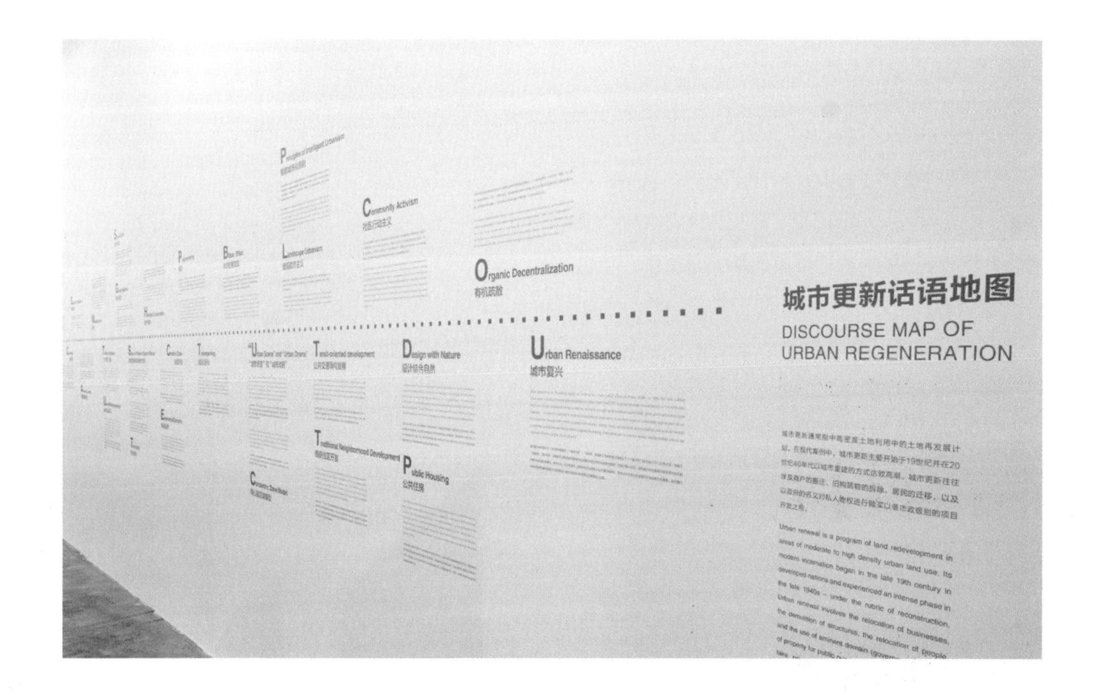


本展位通过梳理城市更新概念发展过程中不断涌现的话语，制作成地图形式，来向观众直观呈现该领域出现过的学术讨论。它全面展现了城市更新在全球范围内不同地域和国别的理论维度思考。

**Emerging discourses through the evolution of urban regeneration concepts, are documented and presented by means of a discourse map, in order to exhibit the academic discussions in related fields. This section reviews the theoretical thinking on urban regeneration in various regions and countries on a global scale.**

更新重要法案

版物

FICANT LAWS AND

CATIONS OF

N REGENERATION

1840s

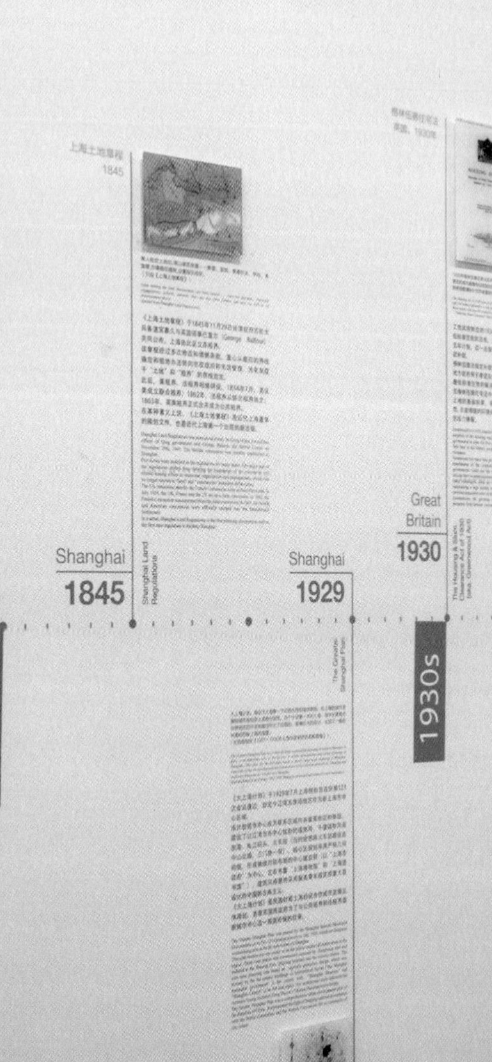

# A-03

## 城市更新重要法案及出版物
## EXEMPLARY LAWS AND PUBLICATIONS OF URBAN REGENERATION

城市更新重要法案与出版物板块着眼于世界范围内的重要法案与出版物研究。此板块研究筛选了 46 个法案，其中包含国际法案 / 出版物 33 项，上海法案 13 项，横向涉及欧洲、美洲及亚洲的一系列重要国家和城市，反映多样的城市更新思路和方法。板块内容涉及著名专家或者机构发起的城市规划宪章、以经济福利保障促进建成区更新、建立层级遗产保护制度来保障城市历史文脉的有机更新、内城更新、都市企业模式推动资产导向型城市更新、知识驱动型经济推动的城市更新、城市扩张政策下的城市更新等。

The section of Exemplary Laws and Publications holds worldwide researches. It selects 46 acts, regulations or publications, of which 33 are international and 13 in Shanghai. A number of countries and cities in Europe, America and Asia are taken in the research scope, reflecting various thoughts and methods in urban regeneration. This section involves charters or regulations raised by experts or institutes, promotion of regeneration under social welfare, maintenance of organic regeneration within historical context by means of establishing leveled preservation system, urban regeneration of inner cities, regeneration led by enterprise mode, etc..

2000
一城九镇
Flood Risk 101
除了风险之外
洪水
能否为城市
带来发展机遇
Can we create a new plaza with old trees?
树挪死，人挪活——我们能不能用老树做个新广场?

# A-04 城市更新经典案例 EXEMPLARY CASES OF URBAN REGENERATION

城市更新经典案例展示板块，包含了 50 个典型案例，其中国外案例 40 个，上海案例 10 个。时间跨度从 1851 年至今，范围涵盖欧洲、美国、东南亚、香港等地区。案例的类型多样，都具备所在时代的典型性，内容包括：典型文化事件促进城市更新；标志物的建造促进城市更新；水岸更新；内城更新；著名机构或者政策引领城市更新；城市扩张过程中带来的城市更新等。此次案例类型和内容齐全，配以文字及图片，为公众全面生动地展示出近一个半世纪以来的世界城市更新进程。

50 Cases (40 abroad and 10 in Shanghai) are included in this section, spanning from 1851 to today, covering Europe, the USA, South East Asia and Hong Kong. These cases include urban regeneration promoted by typical cultural events; urban regeneration enhanced by landmarks; regeneration of waterfronts; regeneration of inner cities; urban regeneration led by institutes or policies; urban regeneration within the expansion of cities. This section is illustrated to provide a general view of the urban regeneration process in the past one century and a half.

城市更新 专家访谈
Urban Regeneration

留学生 拍摄地点：莫干山路
Foreign Student
Location:Moganshan Road
的城市现代化国
are on a so rapid track of urban modernization

# A-05 A-06

## 城市更新专家谈 & 都市漫谈（影片）
## INTERVIEWS OF EXPERTS AND GENERAL PUBLIC ON URBAN REGENERATION

本展位是城市管理者、规划师、建筑师对城市更新的理解与建议的专家访谈，以及对最广大的普通市民的关于对上海城市建设及更新的街头访问。

Interviews with city administrators, planners, architects to discuss the issues of urban regeneration. Interviews with general public for their opinions on urban construction and regeneration.

都市漫谈 URBAN GOSSIP 城里厢嘎讪胡

1. 老城厢里记忆深 Memorable Old Town 大有来头老城厢
2. 传统文化有精髓 Profound Cultural Deposits 老底子文化有味道
3. 改造动迁故事多 Renewal and Removal 房子拆特了搬了老远
4. 商业发展集大成 Complex Commercialization 闹猛地方做生意
5. 同类城市来相较 Shanghai vs. Other Cities 阿拉上海帮伊拉外地
6. 生活交通要便捷 Convenient Lifestyle 生活老方便的
7. 这些地方是楷模 Ideal Plans and Places 阿拉最欢喜的地方
8. 城市更新之我见 Opinions on Urban Regeneration 城市更新阿拉有闲话要讲
9. 畅想未来更精彩 Shanghai Future 希望上海变得更嘎嗲

城市更新
URBAN
REGENERATION

# A-07

## 城市肌理的时间线
## TIMELINE OF URBAN FABRICS

城市肌理有形或无形，只有当其被触动之时产生其作用和意义。哪些要素曾经塑造了城市的肌理？哪些事件一次次地促成了城市肌理的改变？于是，“城市肌理的时间线”研究项目不再是对历史片段的简单回溯和记录，更为重要的是揭示种种城市肌理变迁现象背后的隐形逻辑和联系。

**Whether tangible or not, urban fabrics only manifest themselves when challenged. Which elements composed them in history? Which competitive forces created them? A Timeline of Urban Fabric is more than a chronological retrospect, but also an investigation of the reasons and forces that act upon them.**

**李丹锋、周渐佳 | Li Danfeng, Zhou Jianjia**
冶是建筑工作室 | YeArch Studio

# 回溯

## 历史的承袭与演进

## RETROSPECTIVE

B

## 回溯：历史的承袭与演进

此板块旨在对城市历史保护与更新的相关实践与理论研究进行回溯，呈现地方、国家和全球语境下的新问题与城市策略。展示了中国古城古村落的建筑修复和设施更新的经验成果，国内外历史街区保护与历史建筑改造和文化传承的优秀案例，艺术运作、文化机构与城市互动为动力的城市复兴运动等，展示对当下城市历史保护问题的多维度思考、应对快速城市化的多重策略，以及新的生活方式给旧城更新带来的发展契机。

## RETROSPECTIVE: INHERITANCE AND EVOLUTION OF HISTORY

"Retrospective" reviews practices and theoretical studies on historic conservation and regeneration, while also communicating new problems and urban strategies in local, national, and global contexts. The exhibitions in this section cover the experience of architecture reconstruction and infrastructure renovation of Chinese historic villages, redevelopment projects of historic districts and buildings and their cultural inheritance both in China and worldwide, and the urban renaissance driven by the interaction among artistic actions, cultural institutions, and the city. It demonstrates multi-dimensional considerations for historic conservation, multiple strategies for responding to rapid urbanization, and opportunities brought by the creation of new lifestyles.

社区行动主义
Urbanism
Landscape Urbanism
HAMBURG
HAFEN CITY
QUIOSCOS PONTONES
1/100
HAM-41
98-58-23
HAMBURGO
HAFENCITY '02
PUERTO TRADICIONAL M.1-250
HAF-32
68-45-18

汉堡港，位于易北河入北海口。Hafencity 西侧公共空间，属于邻城历史仓库区南的原海港更新的一部分。纵观历史，这片区域在转变中一直维持其海港和工业用途。高低起伏是码头的常见表现。针对 Hafencity 采用的策略是让滨水空间环城而建，而不仅位于其内。2/3 的新公共空间均可被淹。它们并不是无用之地，而是有意义与活力的。位于高处的城市总不会被水淹。Hafencity 项目的介入是动态与灵活的，是一个以人与水为尺度的多变地景，并将人带至水边。

**Hamburg Port, on River Elbe, flows into the North Sea. The western open spaces of Hafencity are part of the processes of transformation of the former harbor zone south of the historical Speicherstadt, bordering on the inner city. This area has changed continuously throughout its history, keeping both harbor and industrial uses. The alternating ebb and high tides characterize the typical appearance of the port basin. EMBT's strategy in Hafencity is to give water space, rather than contain it, to protect the city. Two-thirds of the new public spaces are floodable. They are not useless remaining spaces, but usable and livable. The city is at a higher level, always protected from the water. Hafencity project is a dynamic and flexible intervention, a changing landscape on a human scale moving concurrently with the floods, attracting the people and bringing them nearer to the water and its moods.**

**米拉莱斯·塔格里亚布 EMBT 建筑事务所 | Miralles Tagliabue EMBT**

米拉莱斯·塔格里亚布 EMBT 是一家国际知名建筑事务所，项目分布欧洲与中国各地，内容包括建筑设计、室内设计、城市设计、历史建筑改造与景观设计。 Miralles Tagliabue EMBT is an international acknowledged architecture studio. Its mature approach to architecture, interior design and facility planning includes educational, commercial, industrial and residential projects, restoration and landscape architecture.

a minimum urban scale, the way how to act in accordance with the new from the research related with studies of the master plan of Havana City which has the certain conclusion of developing the city within itself from its was proposed to identify the forms to do the planning, design, construction an area of intervention in the City of Havana. Now we present a selected , and whose results could be applied in other areas of the city itself and

# B-02 更新中的哈瓦那 HAVANA IN REGENERATION

更新中的哈瓦那展位由三个主题板块组成：1. 哈瓦那城市与建筑在历史中的瞬间；2. 古巴在 20 世纪至今的先锋建筑案例；3. 哈瓦那的可持续发展问题。在第三个板块中，展示了哈瓦那城市修复、可持续的街区更新范例、城市更新中的新旧融合问题，作为对此次展览主题的诠释与呼应。

**The exhibit project has three main parts: a) Moments of the history of the city and its architecture. b) Examples of Cuban Architecture of centuries XX and XXI. c) Havana and urban sustainability. The third part of the exhibition project showed research, works of the academic ambit, related with ways to rehabilitate the city, and its neighborhood for urban sustainability, considering the building rehabilitation and the new constructions as parts of the whole problem.**

**恩里克・费南德兹、张晓春、田唯佳 | Enrique Fernández, Zhang Xiaochun, Tian Weijia**

恩里克・费南德兹 / 哈瓦那建筑学院教授，哈瓦那建筑学院“哈瓦那城市与建筑更新”课题组责任教授

张晓春 / 同济大学建筑城规学院副教授

田唯佳 / 同济大学建筑城规学院助理教授

Enrique Juan de Dios Fernández Figueroa / Professor, Faculty of Architecture of Havana, ISPJAE; Leader of research team in the Faculty of Architecture of Havana: " Urban rehabilitation and its architecture" (RUA).

Zhang Xiaochun / Associate Professor, CAUP, Tongji University

Tian Weijia / Assistant Professor, CAUP, Tongji University

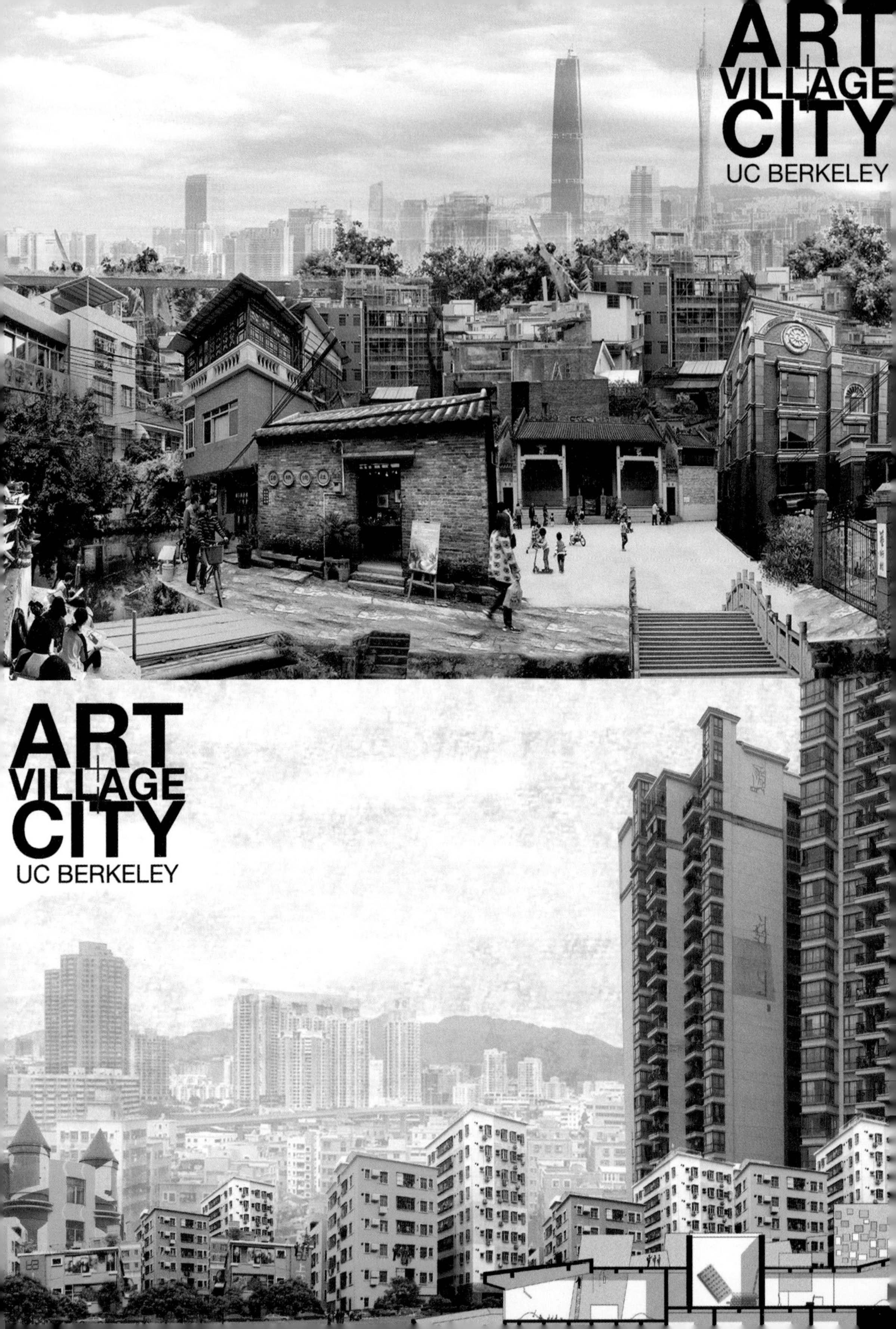
ART
VILLAGE
CITY
UC BERKELEY
ART
VILLAGE
CITY
UC BERKELEY

# B-03 艺术 + 村 + 城市 ART+VILLAGE+CITY

**艺术 + 村 + 城市研究工作室使用跨学科研究方法和协作精神，来研究处于建成环境中的人的境况。本展览是工作室对珠江三角洲艺术家村落的研究作品精选，力求探索广州小洲村和深圳大芬村两个案例，并记录艺术和创造性在城中村和城边村转型中的作用。**

**The Art+Village+City Research Studio utilizes interdisciplinary methods and a collaborative ethos to study the human conditions in the built environment. This exhibition is a selection of the studio's study of artist villages in the Pearl River Delta region. It presents two case studies-Xiaozhou village in Guangzhou and Dafen village in Shenzhen, and documents the role of art and creativity in the transformation of villages-in-the-city and villages-by-the-city.**

**加州大学伯克利分校艺术 + 村 + 城市研究工作室 | Art+Village+City Research Studio at UC Berkeley**

它是由玛格丽特·克劳福德教授和黄韵然教授领导的跨学科研究工作室，由加州大学伯克利学院全球城市人文倡议组赞助。

It is an interdisciplinary research studio led by Margaret Crawford and Winnie Wong, and sponsored by the Global Urban Humanities Initiative at UC Berkeley.

曹杨新村•
社区公共空间城市更新
Urban Renewal of Community Public Space in CaoYang Xincun
参展团队主要人员：
王伟强，Prof. Weiqiang WANG
同济大学建筑与城市规划学院教授
CAUP. TONGJI UNIVERSITY
王 兰，Prof. Lan WANG
同济大学建筑与城市规划学院副教授
CAUP. TONGJI UNIVERSITY
孙施文，Prof. Shiwen SUN
同济大学建筑与城市规划学院教授
CAUP. TONGJI UNIVERSITY
Prof. Margherita Turvani
IUAV. 意大利威尼斯建筑大学教授
Prof. Maria Chiara Tosi
IUAV. 意大利威尼斯建筑大学教授
Prof. Stefano Munarin
IUAV. 意大利威尼斯建筑大学教授
薛玉洁，Yujie XUE
同济大学建筑与城市规划学院研究生
CAUP. TONGJI UNIVERSITY
蒋希冀，Xiji JIANG
同济大学建筑与城市规划学院研究生
CAUP. TONGJI UNIVERSITY
Alex Pellizer
IUAV. 意大利威尼斯建筑大学研究生
Silvia Maroso
IUAV. 意大利威尼斯建筑大学研究生
曹杨新村社区公共空间城市更新
城市结构的多孔性
04
上海 · 普陀 · 曹杨新村
SHANGHAI • PUTUO • CAOYANGXINCUN
CAOYANG
Lanxi Road
06

# B-04

## 曹杨新村社区公共空间城市更新案例实践
## THE URBAN RENEWAL IN PUBLIC SPACE IN CAOYANG XINCUN

本次展览包含两部分内容：1. 设计研究展：结合中外院校长期的研究与设计成果，对曹杨新村问题展开研究，并对未来的发展设想、结构调整、形态演进，尤其是社区环浜公共空间更新改造提出新的构想。这些内容以图版、模型、录像等形式展出。2. 实践案例展：配合街道的要求，同济大学团队针对桂巷路步行商业街进行了更新改造的环境景观设计，这成为社区公共空间改造更新的实践案例展。施工改造的演变过程被做了动态跟踪记录，且将这一过程结合录像视频呈现在了展览中。

**This exhibition contains two parts: 1. The Exhibition of Design Research. With long-term cooperation work between Chinese and foreign institutes, this research studies the problems, the structural adjustment and the aromorphosis of Caoyang Xincun, and then pictures its further development, especially showing a new blueprint on the renewal of community public space around the riverside. All achievements are shown in the way of picture, model and video, and so on. 2. The Exhibition of Practice Research. With the demand of Caoyang Community, team of Tongji University finishes the environmental landscape design on the renewal Guixiang Road as pedestrian street. This is the practice exhibition about the community public space renewal. The construction process was recorded and shown to the public by video.**

**王伟强、王兰、孙施文、图瓦尼、齐亚娜·托西、斯蒂法诺 | Wang Weiqiang, Wang Lan, Sun Shiwen, Margherita Turvani, Maria Chiara Tosi, Munarin Stefano**

王伟强教授、王兰副教授、孙施文教授 / 同济大学建筑与城市规划学院

图瓦尼教授、齐亚娜·托西教授、斯蒂法诺教授 / 意大利威尼斯建筑大学

Prof. Wang Weiqiang, Wang Lan, Sun Shiwen/ College of Architecture and Urban Planning, Tongji University

Prof. Margherita Turvani, Maria Chiara Tosi, Munarin Stefano/ Università IUAV di Venezia

上海｜缩微之城
MICROCITY OF SHANGHAI
Architecture is a picture frame of a city. Microcity of Shanghai presents five selected cases, all of which are with century-long historical changes. This research extracts values of form, functions, symbolism from these cases, and examines the correlation of architecture, urban space and social life. It is to present a history from physical space to city events, collective conceptions and fashion. Audiences can have a glimpse of a general change of Shanghai by viewing the converting scenes in the models.
PARTICIPANTS
SHANGHAI GRAND THEATRE
东正教圣母大堂
SHANGHAI EASTERN ORTHODOX CATHEDRAL
上海工人文化宫
SHANGHAI WORKER'S CULTURAL PALACE

# B-05 上海缩微之城
# MICROCITY OF SHANGHAI

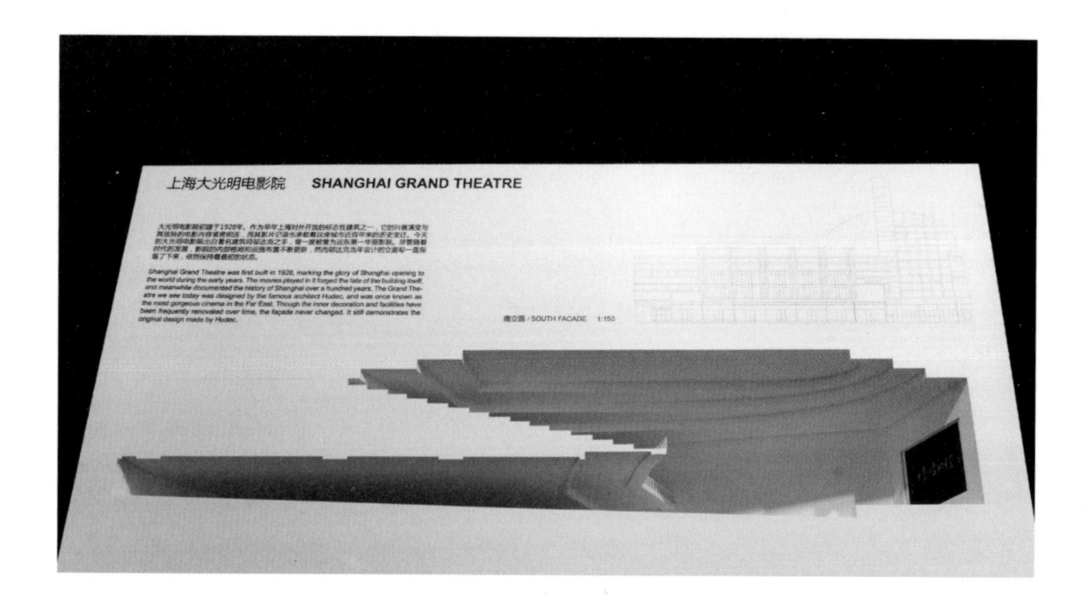


**建筑是瞭望一座城市的景框，本案所选取的五处上海案例经历近百年的历史演变后都成了一座缩微之城。本研究从这五座建筑的更易中区分出形式价值、功能价值、象征价值等层面，放入城市空间与社会生活的变迁中作关联度研究，试图呈现出一段从实体空间到事件、集体观念、风尚的变迁史。本作品呈现在观者面前的，是立体模型配合着场景转换以及视觉音效所表现出来的历史演进过程，观者能管窥上海都市面貌的变迁。**

**Architecture is a picture frame of a city. Microcity of Shanghai presents five selected cases, all of which are with century-long historical changes. This research extracts values of form, functions, symbolism from these cases, and examines the correlation of architecture, urban space and social life. It is to present a history from physical space to city events, collective conceptions and fashion. Audiences can have a glimpse of a general change of Shanghai by viewing the converting scenes in the models.**

**李翔宁、江嘉玮 | Li Xiangning, Jiang Jiawei**

李翔宁 / 同济大学建筑城规学院教授、2016 年哈佛大学设计研究生院访问教授、国际建筑评论家委员会委员

江嘉玮 / 同济大学建筑城规学院博士候选人、耶鲁大学访问研究员

Li Xiangning / Professor, CAUP, Tongji University; Guest Professor in 2016, GSD, Harvard University; Member of International Committee of Architectural Critics

Jiang Jiawei / PhD Candidate, CAUP, Tongji University; Visiting Assistant in Research, Yale University

# B-06

## 荷兰国立博物馆新馆与巴塞尔中央火车站更新项目
## PROJECT OF RIJKSMUSEUM IN AMSTERDAM AND CENTRAL RAILWAY STATION IN BASEL

**本展位选取的两个案例展现了克鲁兹和奥尔蒂斯事务所关注的老城区更新中建筑与城市空间的关系：1. 荷兰国立博物馆大楼是位于阿姆斯特丹的一栋修建于 19 世纪的建筑物，长久以来它担任着衔接老城区与南部新城区的文化纽带这一重要角色；2. 巴塞尔火车站项目将重新审视它作为城市基础设施该扮演的角色，既作为城市的迎客之门，也作为周边区域的连接点。**

**Two selected cases exhibit the concerns of Cruz y Ortiz Arquitectos in the relations of in architecture and urban space within the regeneration process. 1. The Rijksmuseum in Amsterdam, which is located in a building constructed in the 19th century, has played the role of linking the ancient downtown and the new southern district for a long time. 2. Central railway station project in Basel re-examines its role of urban infrastructure, both as a city portal and as a juncture of surroundings.**

**克鲁兹和奥尔蒂斯建筑师事务所 | Cruz y Ortiz Arquitectos**

总部设在西班牙塞维利亚市，所接触的建筑设计的范围非常广泛，从住宅、学校到体育馆、火车站、国家博物馆这类的大型基础设施都有涉猎。 It is headquartered in the city of Seville in Spain. It has developed projects in a wide range of fields and at very different scales: from private houses to schools, stadiums, railway stations or national museums.

# B-07 陆家嘴提喻法 SYNECDOCHE LUJIAZUI

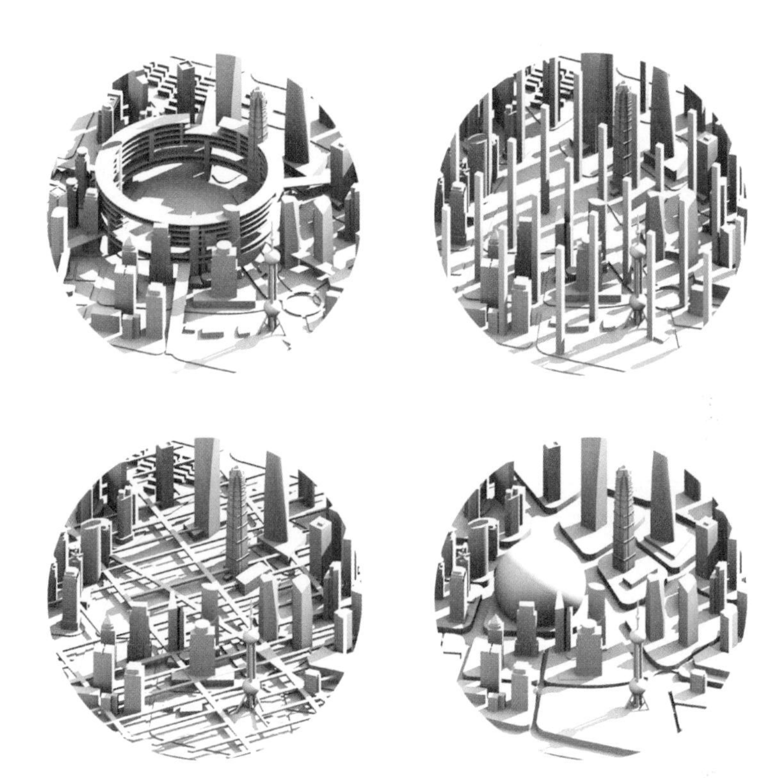

无论是建筑师、规划者、政策制定者，还是生活、工作、旅行于此的人们，都在以各自的方式或真切或虚幻地重塑着这座城市。如同卡尔维诺笔下的威尼斯，理解城市的困境不在于面目不清，而在于其复杂性与多变性。提喻法提供了在个体与集体、局部与整体、抽象与具象、当下与未来之间重读陆家嘴的新途径，无数座城市由此诞生。

**Architects, planners, policy makers and people who live, work, travel in constantly reshape the city in their own way, physically or virtually. Just like Venice in Calvino's world, vagueness never obstructs the reading of a city, instead, its complexity and ever-changing nature does. Synecdoche explores a new way to capture the essence of Lujiazui between the individual and the collective, the partial and the whole, the abstract and the concrete, the realistic and the futuristic, thus myriad of cities birthed.**

**李丹锋 | Li Danfeng**

冶是建筑工作室主持建筑师、同济大学建筑城规学院博士候选人

Phd Candidate; Principal Architect, YeArch Studio

上海 棋院

# B-08 上海城市更新实践 PRACTICE OF SHANGHAI URBAN REGENERATION

此展览共涵盖了 12 个上海在地的城市更新案例。建筑师以各自独特的视角，用不同的建筑手法延续了城市有机体的生命力，也为城市生活注入了新的活力。其中六个项目围绕“后工业城市复兴”的主题，将旺盛的工业时代过去后被“遗弃”的城市组件进行部分的改造，赋予其新的使命；另外六个项目则以“社区共生”为主旨，将被人们忽略的“灰色地带”重新带入现代的社区生活，甚至成为城市社区中富有活力的一个重要组成部分。通过这 12 个策略各异的作品，引发专业者和市民能重新审视城市更新的方法与结果，并共同思考未来城市的可能性。

This exhibition introduces 12 recent projects of urban regeneration in Shanghai, whereby architects used different strategies to revitalize our urban spaces and to enhance the life of ordinary citizens. Six of these projects centered around the theme of "Post-industrial Urbanism", transforming abandoned industrial buildings as architectural components for new urban programs. The other six emphasized on notion "Symbiotic Community" by intervening in previously overlooked "Grey Field" within the neighborhood, and bringing new energy to the neighborhood. Through these 12 unique projects, both professionals and the general public will have a chance to re-examine the methods and results of urban regeneration in Shanghai, and to re-imagine the possibilities of the future of our city.

**刘宇扬建筑事务所、麟和建筑、亘建筑 / 孔锐 + 范蓓蕾、直造、大舍 + 旭可、梓耘斋、阿科米星、原作 / 张姿 + 章明、创盟国际、山水秀、曾群 / TJAD ATELIER LIU YUYANG ARCHITECTS, ATELIER L+, genarchitects/Kong Rui + Fan Beilei, Naturalbuild, ATELIER DESHAUS + ATELIER XUK, TM Studio, ATELIER ARCHMIXING, Original Design Studio/ZHANG ZI + ZHANG MING, ARCHI-UNION, Scenic Architecture, Zeng Qun/ TJAD**

**滨江爱特公园 / Riverfront Aite Park**
刘宇扬

**环境建构：上海虹桥、银河宾馆改造 / Environmental Architectonics: The Renovation of Hongqiao and Yinhe Hotel, Shanghai**
李麟学

**同和凤城园 / Tohee Fengcheng Park**
孔锐 + 范蓓蕾（亘建筑）

**外马路 1178 号创意办公改造 / 1178 Waima Road Warehouse Renovation**
水雁飞、马圆融、苏亦奇（直造）

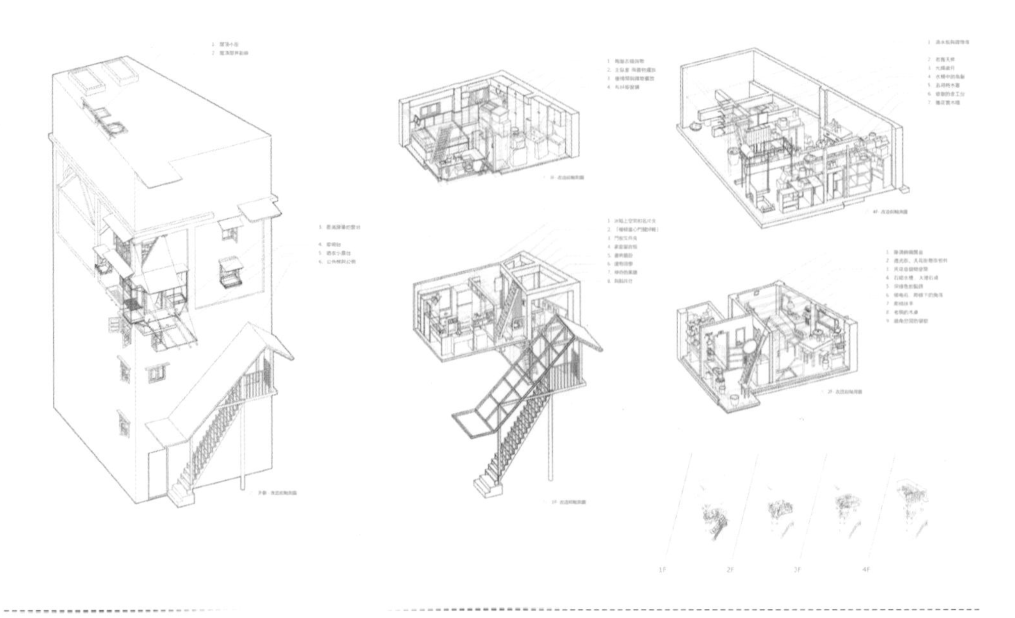

**花园坊 水塔住宅改建项目 / Garden Lan. Tower House Renovation**

柳亦春、刘可南（大舍 + 旭可）

**集装箱售楼处 / Container Sales Office**

刘可南、张旭、陆均、徐亮（旭可）

**韩天衡美术馆 / Han Tianheng Art Museum**
童明

**陈化成纪念馆移建改造 / New Chen Huacheng's Memorial Hall**
阿科米星

**原作设计工作室改造 / Original Design Studio Renovation**
张姿 + 章明 ( 原作 )

**宝山张庙工人新村社区公园设计 / Zhangmiao Exercise Park, Baoshan**
袁烽

**胜利街居委会和老年人日托站 / Victory Street Neighborhood Committee and Oldies Daily-Care Centre**
祝晓峰

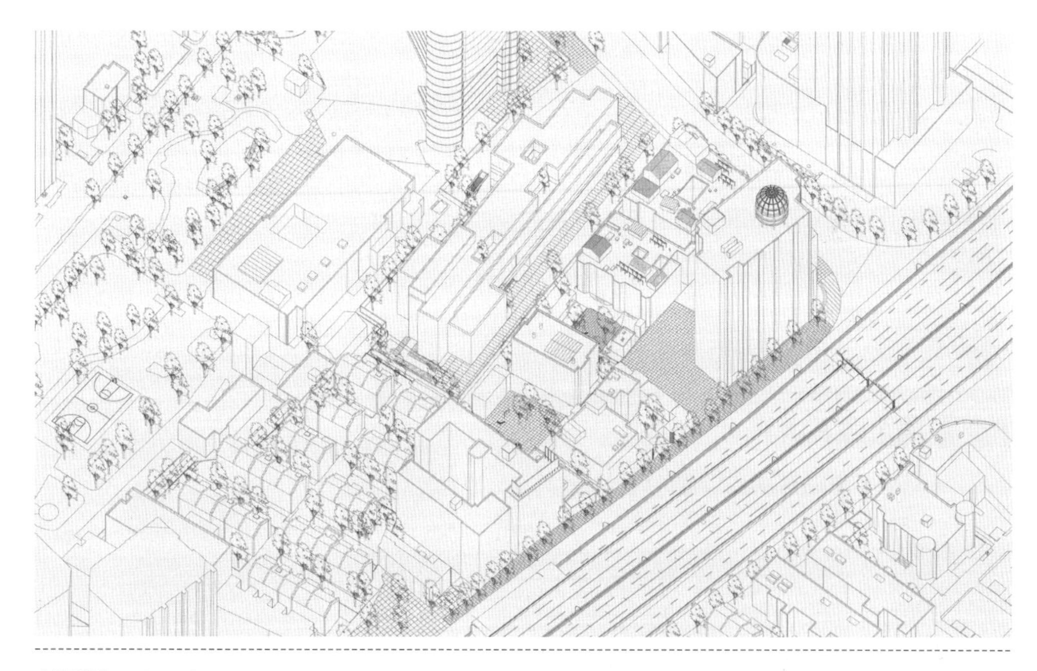

**上海棋院 / Shanghai Qiyuan**
曾群

CUSTOM NARRATIVES
NATURE IMAGINED
CUSTOM ASSEMBLED
TECTONIC STORIES
MENTALIZED

# B-09 定制 CUSTOM MADE

在挪威建筑和城市发展中，自然化的传统是指习俗与定制之间的相互作用。本次展览是围绕概念化的建筑构成核心要素：墙与书。这面墙展示了建筑在学院派思想指导下被建造；这本书展示了建筑如何被书写与呈现，揭露了建筑是如何被理解、探讨和论证建筑的方式。这种二元论可以被分别定义为“制造的传统”，指建筑实践，以及“传统的制造”，指建筑的争辩与讨论。“定制”，不是意在为挪威建筑和城市发展置入新的标准，而是旨在呈现一种衍生性与多样性。本次展览展示了挪威建筑文化生态，这种传统不是单一与一致的，而是对定制先例的集合，这种集合既是建造方面也是书面理论的。

**Naturalizing tradition refers to the interplay between convention and customization in Norwegian architecture and urban development. The exhibition is conceptualized around the central elements constituting architecture: The Wall and The Book. The Wall shows architecture as built within a school of thought. The Book shows architecture as written and represented, referring to how architecture is understood, discussed and legitimized. This dichotomy can be defined as respectively "the tradition of production", meaning architectural practice, and "the production of tradition", referring to architectural debates and discourses. Custom Made is not about projecting a new canon onto Norwegian architecture and urban development. It is about proliferation and multitude. This exhibition exposes the ecology of architectural culture in Norway, not as a singular and unanimous tradition, but as a collection of customized precedents, both the built and the written ones.**

**米尔扎 · 穆也兹诺维柯、哈尔沃 · 韦德 · 埃乐弗森、史洋 | Mirza Mujezinovic, Halvor Weider Ellefsen, Shi Yang**

米尔扎 · 穆也兹诺维柯 / MALARCHITECTURE 事务所主持建筑师

哈尔沃 · 韦德· 埃乐弗森 / 建筑师，挪威奥斯陆建筑与设计学院在读博士，助理教授

史洋 / 荷兰注册建筑师，hyperSity 事务所主持建筑师，现任教于中央美术学院建筑学院

Mirza Mujezinovic / Principal Architect, MALARCHITECTURE

Halvor Weider Ellefsen / Architect, Ph.D research fellow and assistant professor at The Oslo School of Architecture and Design (AHO)

Shi Yang / Dutch licensed architect (SBA), principal of hyperSity office and assistant professor at Central Academy of Fine Arts (CAFA)

# B-10 闹海 “NOW SHANGHAI”

本展位关注上海与超级大都市中身份认同的争议的空间概念。这项研究对如下课题展开思辨性研究：上海的城市空间形态重构，身份认同的形成与演变过程，历史的作用与未来的展望，审美、原真与社会意向，以及诸如地图、图像、文字和叙事等特殊的再现模式如何塑造了城市的身份认同以及我们的理解等话题。继关于上海的历史、电影、文学、建筑和规划的一年课程之后，洛加大都市人文学项目的跨学科团队赴上海进行了四个基地的调查与拍摄，从寓言到纪录片等不同类型中捕捉到了复杂的叙事。同时，它还以一系列信息密集的空间图解和地图电影化地再现了上海，展现了都市人文学者对上海的过去、当下和未来的多维视角。

**This project focuses on the city of Shanghai and the concept of contested spaces of identification. Issues addressed include spatial and morphological restructuring of the city; forms and processes of identification; uses of the past and visions of the future; aesthetics, authenticity, and the social imaginary; and how particular modes of representation like maps, images, texts, and narratives shape urban identities and our understanding of them. Following a year's course of study about the history, film, literature, architecture, and planning of Shanghai, trans-disciplinary teams of Urban Humanities Initiative of UCLA traveled to the city to explore and film in-situ. Their films capture complex narratives in genres ranging from fable to social documentary. By representing Shanghai cinematically, a series of densely rendered spatial diagrams and maps give multi-dimensions to the urban humanists' view of Shanghai in the past, present and future.**

**达娜 · 科夫 | Dana Cuff**

梅隆基金资助的城市人文学项目首席研究员、洛加大建筑学院城市实验室负责人。Lead Principal Investigator for the Urban Humanities Initiative funded by the Mellon Foundation, and Director of AUD’ s think tank, city LAB.

THERE IS
PERIPHE
无边界
数据检索
DATA
RETRIEVAL
LANDSCAPE
LAB
景观实验室
激进的当代性
RADICAL
CONTEMPORANEITY
SANT BOI OLD TOWN
CAN GINESTAR PARK

# B-11 巴塞罗那 25 年城市更新经验
# 25 YEARS OF URBAN REMODELING OF THE METROPOLITAN AREA OF BARCELONA

巴塞罗那被广泛认为是世界上所有城市的一个城市更新典范，特别是针对市区内公共空间的品质提升。展览从巴塞罗那市政府主导下建成的 1000 个不同尺度与特点的城市公共空间更新项目中，选取了部分优秀案例，来展示近 25 年来巴塞罗那城市更新经验。

**Barcelona is widely regarded as a role model for cities all over the world wanting to leap forward and establish qualified public space at the heart of their transformations. The exhibition shows a selection of projects realized by the administrative body governing the metropolitan area of Barcelona, AMB, out of the 1000 recorded in the past 25 years, representative of different scales, situations and character.**


**巴塞罗那政府、恩里克 · 玛西普 | Metropolitan Area of Barcelona, Enric Massip**

巴塞罗那政府

恩里克 · 玛西普 / EMBA 建筑师事务所主持建筑师

AMB - Metropolitan Area of Barcelona

Enric Massip / Principal of EMBA Architects

# B-12 北京城市更新实践 PRACTICE OF BEIJING URBAN REGENERATION

爵士乐中心（前门 23 号地下空间改造）
Chiasmus/ 柯卫

此次北京城市更新展区由 11 位建筑师的 11 个作品组成，从 11 个角度构成了北京这一古老城市更新的当代群像。其中五位建筑师聚焦于"胡同"，借精确的单体改造与细腻有分寸的手法实现精巧迷人的空间与独特的当代功能置入；另外六位建筑师面对北京复杂的既有建筑与浓厚的历史文脉进行当代性文化性的改造，试图激活建筑的潜在价值并对所在区域产生积极高品质的影响。望这 11 位建筑师的作品中对当代性的"新"与城市文脉的"旧"的关联与对立的思索，能唤醒对中国当下"城市更新"更为审慎克制、有责任感与进取心的态度。

**This year's Beijing City Renewal Exhibition comprises of 11 works by 11 architects, providing 11 contemporary perspectives on what renewing this historic city may mean. Five architects placed their focus on the Hutongs of Beijing, while synthesizing modern program with existing but delicate locales, they've created exquisite spaces through precise and delicate insertions or modifications. The other six architects worked within the complex, rich underlying culture and history of Beijing itself, activating the underlying potential of existing structures in an attempt to rejuvenate surrounding areas. These 11 works are attitudes towards the juxtaposition of "new" and "old" within this city. They are an initiation into a more responsible and responsive intervention in this city that begins with the micro urban environment.**

**Chiasmus/ 柯卫、大章建筑 / 戴璞、标准营造 / 张轲、META/ 王硕、直向 / 董功、OPEN/ 李虎、TAO · 迹 / 华黎、何崴、张利、朱锫、徐甜甜**

Chiasmus/James Wei Ke, Daipu Architects / Dai Pu, Standarchitecture/Zhang Ke, META/Wang Shuo, Vector Architecture/Dong Gong, OPEN/Li Hu, TAO/Hua Li, He Wei, Zhang Li, Zhu Pei, Xu Tiantian

**树美术馆**

戴璞

**被拉长的工业盒子（北京某禅茶会所）**

何崴及其团队

**上海苏河湾竖向艺术中心**
标准营造

**微杂院　北京胡同更新**
标准营造

**箭厂胡同文创空间**
王硕 | META- 工作室

**西海边的院子**
王硕 | META- 工作室

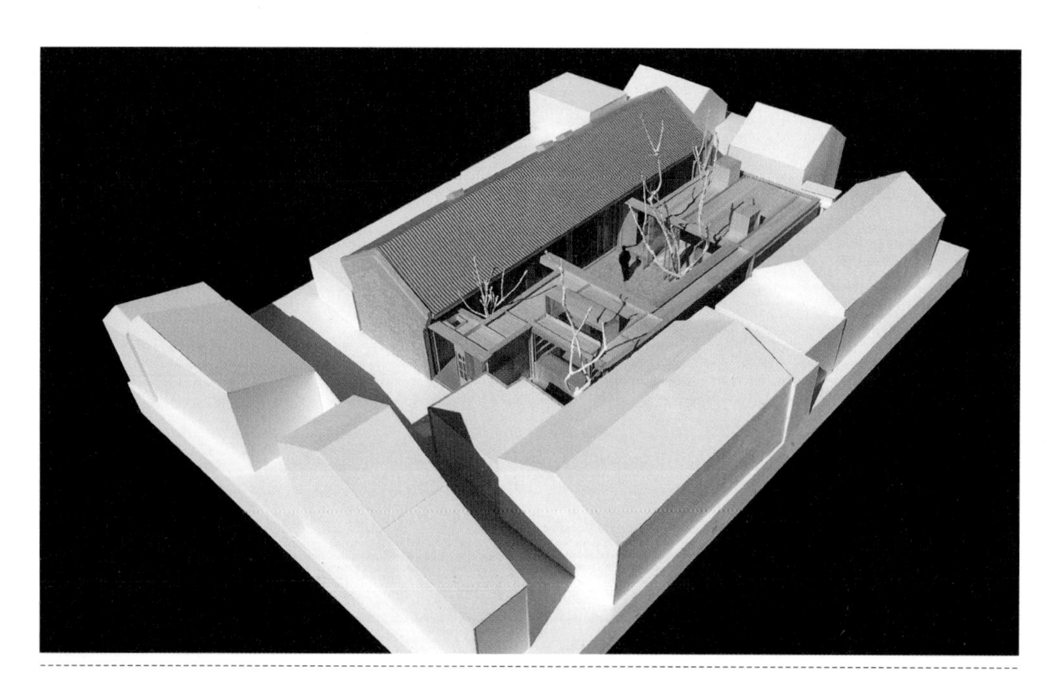

**新杂院**

董功 / 直向

**方家胡同 46 号院恒温车间改造**

OPEN 建筑事务所

**四分院**

TAO 迹·建筑

**南锣鼓巷游客到访中心及商业**

张利

**民生现代美术馆**

朱锫

**宋庄艺术公社**

徐甜甜

# B-13 城市更新回望上海 URBAN GENERATION AND RETROSPECT OF SHANGHAI

展览利用悬挂的方式，将学生的砖雕、木刻作品展示在公共面前，意在倡导城市更新过程中对民间传统工艺的再创造，让文化遗产在城市化的推进中获得新生。通过穿插交叠的方式紧密悬挂木刻与砖雕作品，隐喻民间工艺丰富的文化内涵和其在历史上曾经的辉煌与蓬勃，向两侧渐疏排列透出现代城市背景，暗示了城市过快的更新进程对传统文化的冲击。体验区提供参观者当场体验砖雕、版画印刷过程，使参观者对民间传统工艺有更进一步的了解。

**The exhibition displays students' brick carving and wood engraving works to the public via suspension mode, aiming at advocating recreation on traditional crafts in urban regeneration process, so as to facilitate the renaissance of cultural relics in the boosting of urbanization. In addition, traditional cultural experience area is arranged for visitors to experience the process of brick carving and woodcut print, which can increase exhibition interactivity and enable visitors to have further understanding on traditional crafts.**

**阴佳、田唯佳 | Yin Jia, Tian Weijia**

阴佳 / 同济大学建筑城规学院，教授

田唯佳 / 同济大学建筑城规学院，助理教授

倪凌、徐语键、张佳择、张晶轩等同济大学建筑与城市规划学院学生

Yin Jia / Professor, CAUP, Tongji University

Tian Weijia / Assistant Professor, CAUP, Tongji University

Student of Tongji University: Ni Ling, Xu Yujian, Zhang Jiaze, Zhang Jingxuan, etc.

## 成都宽窄巷子历史文化保护区的复兴

宽窄巷子是成都市最重要的历史文化保护区，它代表了成都市“两江环抱，三城相重”的城市格局，代表了2000 年少城文化和300 年满城文化的传承，代表了北方胡同与川西民居的结合，代表了成都平原的市井生活特征。

自2003 年保护工程启动伊始就备受世人瞩目。核心保护区位于成都市中心，占地约6.6h㎡，原有民居建筑面积约61,300㎡ 共计944 户居民。其中传统建筑院落45 处，占总用地的70%。2008 年建成后总面积约61,785㎡，容积率保持不变。

在8 年的实践过程中，设计重点研究了历史街区的保护方法与策略，规划与建筑设计中的消防问题，木结构和砖木结构保护加固的技术措施，传统木构的承重与抗震设计，传统民居建筑的保温节能设计等等。特别是在现行规范无法完全指导的情况下，既要保证传统建筑形式与风貌，还要保证现代使用功能与安全舒适。经过不懈的努力与坚持，设计团队完成了从研究、规划、测绘、方案、建筑施工图、景观、商业规划到室内设计指导全过程的设计，以更高的社会责任感和更强的历史与人文意识解决了当代中国建筑问题。

项目建成后，形成了使用者与建成空间之间的良性互动。不仅延续了原有的街巷系统与院落肌理，更延续了原住民社区的生活传统，名副其实地成为了近年来中国不多的历史街区有机更新项目中令人称道的案例，将为其他类似项目提供长久的参考与灵感。

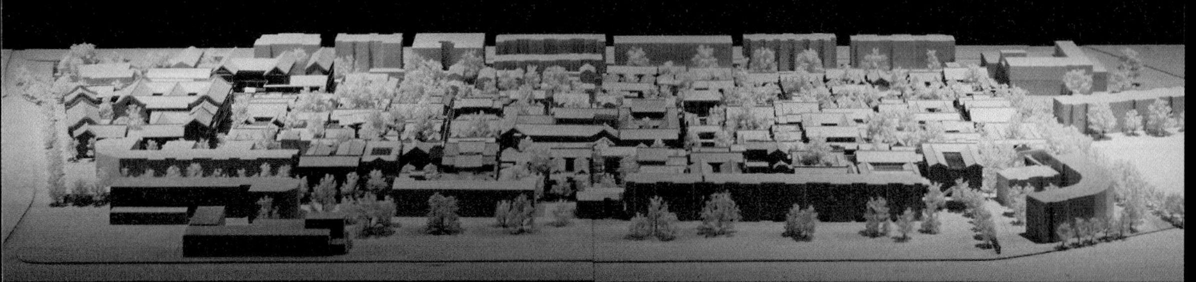

# B-14 宽窄巷子 KUANZHAIXIANGZI

宽窄巷子是成都市最重要的历史文化保护区，历史文化内涵丰厚。核心保护区位于成都市中心，占地约 6.6 公顷，原有民居建筑面积约 61 300 平方米，共计 944 户居民。其中传统建筑院落 45 处，占总用地的 70%。2008 年建成后总面积约 61 785 平方米，容积率保持不变。研究宽窄巷子的历史文化就会发现，它代表了成都市"两江环抱，三城相重"的城市格局，代表了 2000 年少城文化和 300 年满城文化的传承，代表了北方胡同与川西民居的结合，代表了成都平原的市井生活特征。

**Chengdu KuanZhaiXiangZi alley is the most important historical and cultural conservation area with rich historical and cultural significance. The core protection zone at the center of Chengdu covers about 6.6 hectares, with the original residential construction area of about 61,300 square meters and a total of 944 residents. There are 45 traditional courtyards, occupying 70% of the total land. After the 2008 completion of a land about 61,785 square meters, floor area ratio remains unchanged. The research of KuanZhaiXiang-Zi alley of history and culture shows it represents the urban pattern of "being surrounded by two rivers, with three cities encountering" in Chengdu. It presents 2000 years of Shaocheng City Culture and heritage of 300 years of Mancheng City, a combination of alleys in Northern China and vernacular houses in Western Sichuan.**

**北京华清安地建筑设计事务所有限公司 | An-Design Architects Co., Ltd.**

作为清华大学建筑学院建筑设计实践的基地，"产学研"三结合的实践平台，华清安地在工业遗产与保护再利用、历史文化街区等学术领域进行了积极探索，建立了强大的学术研究及设计团队。历年来获得 80 余项部级以上建筑设计奖励，建成了一批有影响的设计作品，充分展现了清华建筑设计的风采和水平。It is an architectural design practice base for Architecture School of Tsinghua University. During its development, An-Design Architects has gradually formed several industrial directions, and carried out studies and explorations in several academic fields, such as the conservation and reuse of industrial heritage, historical and cultural blocks.

社会凝聚力
SOCIAL
COHES

# B-15

## 发现——巴塞罗那建筑学院文献展示
## UNCOVERED — A VISUAL ESSAY FROM ETSAB ARCHIVES

本展位以规则网格展出超过 700 幅等大的文献。每一个文献都带有单独的照片、图解、图片与文字，遴选自 140 年来巴塞罗那建筑学院中的个人或集体。一个有轨带轮移梯让观众可以自由上下，更近距离地观赏每一份档案。本展位在展期之内随机改变整个展墙，因为有些档案明显会比另外的更受观众欢迎，它们与观众之间的关系变得不一样了。

**Our exhibition displays more than 700 equally sized documents arranged within a regular grid. Each document (maximum size DINA4) will contain a single photo, drawing, image or text produced by a selection of ETSAB's individuals and collectives during its 140 years of existence. A specifically designed ladder with wheels will be attached to a rail alongside the wall, allowing visitors to move it sideways, in order to ascend, get closer to every image to properly see its content, and take a copy with them if they want to. This last action also randomly transformed the whole texture and surface of the display during the exhibition, when some documents were evidently taken more than others, physically showing how they resonate differently with the public.**

**巴塞罗那建筑学院 | Escola Tècnica Superior d'Arquitectura de Barcelona**

# 映射

## 城市 / 乡村两生记

## REFLECTION

映射：城市 / 乡村两生记

当下中国的都市实践的视野不再囿于城市。在政策与公共事件背后，乡土实践越来越得到青年建筑师和学者的关注。通过展示 2000 年以来一批优秀中外建筑师的乡土实践、研究学者对未来乡村发展的类型与原型研究，以及艺术家为延续乡村社区文化而尝试的社会性介入，"映射"板块旨在回应过度城市化与失衡的城乡关系，思考乡村哲学，折射对都市建设的反思，并探讨未来城乡融合互动实践在政治经济层面展开的可能性。

REFLECTION: URBAN/ RURAL DUAL LIFE

The vision of urban practice in contemporary China is no longer constrained by the city boundary. In the context of relevant policies and public events, the practices of rural areas have attracted increasing attention from young architects and scholars. "Reflection" will respond to excessive urbanization and the unbalanced urban-rural relationship, rethinking rural philosophy, reflecting on city construction, and investigating the possibilities of the interactive and reciprocal practice between the urban and the rural on both political and economic levels.

# C-01 上物溪北度假农舍 PLACID MOGAN COTTAGE

上物溪北民宿酒店坐落在浙北山谷的茶山竹海中。位于小溪北岸的狭长用地，东西方向长约百米，南北方向宽不足三十米。基地上原有一座乡村小学，校舍靠北侧一字排开，空出南侧场地。设计师沿用原有的场地关系，并将建筑化整为零后结合现有场地高差分散布置，在保证酒店每个客房都拥有南向景观与日照条件的同时，兼顾酒店客房的私密性、服务动线的隐蔽性和便捷性，以及外部场地的丰富性。原有小学校舍的青砖黛瓦双坡顶，也成为延续场地记忆的起点。

**The cottage is located in a bamboo and tea plant valley in the north of Zhejiang province.With an east-west length of more than 100 meters and a north-south width of less than 30 meters, the narrow site is on the northern bank of a brook. There was once a primary school on the north boundary of this site, and the rest southern part was a playground. Based on the site, the architects distributed the mass volume into small components, and then spread them onto the ground according to different elevations of the site. The design intends to maximize the view of the landscape in the south and fulfill the requirements of sunshine, with the consideration of the privacy, imperceptibility and flexibility of the service circulation, and diversity of the exterior. The double pitch black tile roof of the primary school is where to start to follow the memory of the site.**

**亘建筑 | Genarchitects**

亘建筑事务所由孔锐和范蓓蕾于 2012 年在上海创立。孔锐在南京大学获得建筑学硕士学位，曾长期供职于德国 gmp 建筑事务所。范蓓蕾拥有同济大学建筑学硕士学位和德国柏林工业大学城市设计硕士学位。 Genarchitects was founded by Kong Rui and Fan Beilei in Shanghai, 2012. Kong Rui received his Master Degree of architecture from graduate school of architecture, Nanjing University. Fan Beilei holds a Master Degree of architecture from Tongji University, and a Master Degree of Urban Design from Technical University of Berlin.

前童木

平田村隶属浙江省丽水市松阳县四都乡，这个公益项目任务是将村口几栋破损严重荒废闲置的夯土村舍改造成为新的村民中心，同时成为对外展示乡土农耕文明和传统手工艺文化的窗口。平田农耕博物馆及手工作坊由旧夯土房屋改造而成。设计师通过寻找建筑原有的秩序，基本保留了原有建筑风格、形式，使之得以与周围环境保持和谐统一。浙江省西南部的松阳县是中国传统村落保护发展示范县，松阳盛产的竹子作为建构材料，可以减少对茶园生态环境的影响，用料环保，结构轻盈简洁，震害轻，施工速度快。整体采用一系列单体亭子和平台，贴近茶田并自然围合出小庭院。

**Pingtian village is located in Sidu township, Songyang County, Lishui City, Zhejiang Province.**
**This project aims to serve the common good by rebuilding several obsolete rammed earth houses into a new activity center for village people and a window to present the local agricultural culture and traditional craft techniques. The designers sought for the logic of existing buildings and harmony between the new project and the old buildings through the retention of existing architectural style and form. Songyang County, located in the south of Zhejiang Province, is the exemplary county for Development and Protection of China Traditional Villages. The local bamboo is used as construction material which reduces the negative impact upon the local environment. The pavilion creates a space for small garden near the tea plantation by utilizing a series of single pavilion and platform.**

**徐甜甜 / 北京多维度建筑咨询公司 | Xu Tiantian/DnA_ Design and Architecture**
自 2004 年起至今在北京主持 DnA_Design and Architecture 建筑设计事务所，中科院建筑设计和研究中心客座教授。
Founding architect of DnA_Design and Architecture since 2004. Visiting professor of Institute of Architectural Design and Research Center.

# C-03 武夷山竹筏育制场 WUYISHAN BAMBOO RAFT FACTORY

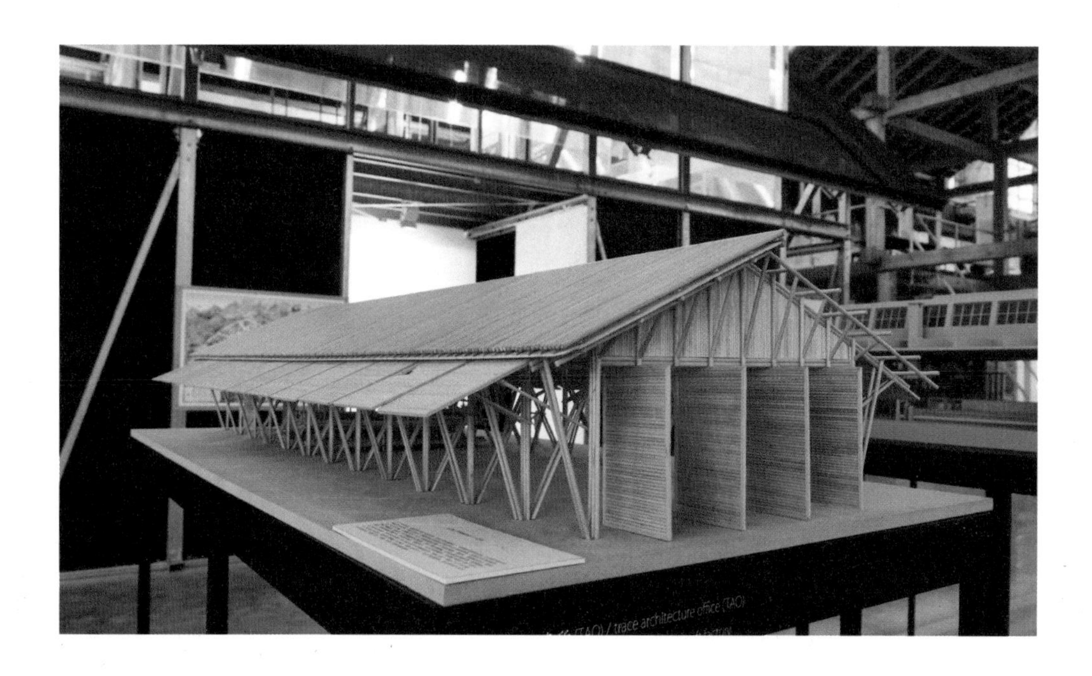


武夷山竹筏育制场是武夷山九曲溪旅游漂流用竹排的储存及制作工厂，由竹子储存仓库、竹排制作车间、办公及宿舍楼这三栋建筑及其围合的庭院组成。建筑的布局与朝向结合地形、风向考虑，建筑主体采用素混凝土结构和混凝土砌块外墙，屋面采用水泥瓦，竹、木作为遮阳、门窗等元素出现。项目的工业厂房性质决定了建筑摒弃形式上的多余，而在建构上采用最基本的元素，并尽可能呈现其构造逻辑，在营造工业建筑朴素美学的同时获得经济性。

**Located atop a plateau in the rural Xingcun village, Wuyishan Bamboo Raft Factory is a building complex for the manufacture and storage of bamboo rafts used to sail the Nine Bend River in Wuyi Mountain for tourism. The site layout responds to the distinct programmatic, topographical, and climatic requirements. The main structure is constructed with in-situ concrete, hollow concrete blocks, and cement tiles. Bamboo and wood are used for sun-shading, doors, and windows. By using the most basic elements, the architecture reveals its structural and material logic; and achieves a simple but not simpler aesthetics with a limited budget.**

**华黎 | Hua Li**

迹・建筑事务所 (TAO) 创始人及主持建筑师，毕业于清华大学和耶鲁大学，获建筑学硕士学位。Hua Li received his B. Arch. and M. Arch. from Tsinghua and Yale University. He is the founder and principal of TAO office.

# C-04 西河粮油博物馆 XIHE CEREALS AND OILS MUSEUM

项目位于中国河南省信阳市新县西河村，旨在通过好的设计，修复环境，激发活力，重振经济。建设资金来自于政府补贴和村民组成的合作社，建设队伍主要由留守村民组成，并由农民自己经营。这个项目更像是一个公共事件，它促使村民重新团结起来，为村庄振兴出力。同时，通过劳动和产业复兴，村落也可以重新找到自我。完成后的西河粮油博物馆及村民活动中心包括一座微型博物馆、一个村民活动中心和一个餐厅。这里将成为西河村新的公共场所，激发村庄的活力；同时它也将拉动村庄经济，成为农村产业重塑的抓手。

**This project is situated in Xihe Village, Xin County, Xinyang city, Henan province, China. The aim of the project is to restore the environments, vitalize the area and revive the economy of the village through a good design.The project was funded by the government and the villagers' cooperative party. The construction was mainly done by the left-behind villagers, and managed by the villagers. This project is more like a public event. It calls for the villagers to stand together for the revitalization of village. Meanwhile, through collective labor and industrial recovery, and the village can also find its position again. The completed Xihe Cereals and Oils Museum and villagers' activity center includes a micro museum, a villagers' activity center and a restaurant. It will be a new public place for Xihe village, vitalize the whole area, and meanwhile revive the economy of village and the rural industry.**

**何崴 | He Wei**

1973 年生于北京，清华大学建筑学院获建筑学学士，德国斯图加特大学获建筑与城市规划硕士，中央美术学院建筑学院获设计艺术学博士。Born in 1973, Beijing, China, He is B. Arch. School of Architecture, Tsinghua University (Beijing, China), Dipl. Ing. -Architecture, University Stuttgart (Germany) and Ph. D. of School of Architecture, Central Academy of Fine Arts (Beijing, China).

# C-05

## 作为城市的乡村
## THE COUNTRYSIDE AS A CITY

**30 年来以城市为中心的发展策略让中国产生了独特的城乡差异。未来城市化率不会再来自大城市的扩张，而主要源于小城市和村镇的增长。中国的农村城市化不应被误解为英国的花园城市，或美国由私有制和汽车造成的郊区化，也应避免土地开发等于高密度城市的假设。不过，农村城市化也是完全保留乡村的全部印记，让设计限于提供卫生和基础设施。本课程的重点在于如何想象一个自给自足、充满经济活力的农村，同时创造文化价值，挑战城乡差异，让城市成为一个平等有包容性的概念。**

**More than three decades' urban-centric developments has entrenched China's unique rural-urban gap. The increase in urbanization ratio in the future will not come from the further expansion of large cities but will instead be focused on the growth of rural towns and small cities.China's rural urbanization should not be confused with the creation of picturesque garden cities in Britain or the sub-urbanization of the United States that was made possible by private land ownership and the automobile. It is likewise key to avoid the assumption that any form of development must be anchored by a dense urban center, wherever that may be. On the other hand, rural urbanization should also shy away from the uncritical position of preserving every surviving fragment of a village, reducing design action to the mere provision and upgrading of infrastructure and sanitation. The challenge for the studio was to imagine a self-sufficient place able to support a dynamic economy in the countryside, providing cultural and intellectual stimulation and offering a respite from the inequalities and divisions that plague the developmental city; in other words, to imagine the city as a space of equal and plural coexistence.**

**克里斯托弗 · 李志明 | Christopher Lee**

克里斯托弗 · 李志明是思锐建筑事务所（伦敦、孟买和北京办公室）的共同创始人和负责人。他同时任哈佛大学设计学院城市设计实践副教授。 Christopher Lee is the co-founder and principal of Serie Architects London, Mumbai and Beijing. He is Associate Professor in Practice of Urban Design at Harvard University's Graduate School of Design.

乡建宣言
——乡村自然营造法则
RURAL MANIFESTO
the natural construction principles

# C-06 竹林构 BAMBOO FOREST CONSTRUCTION

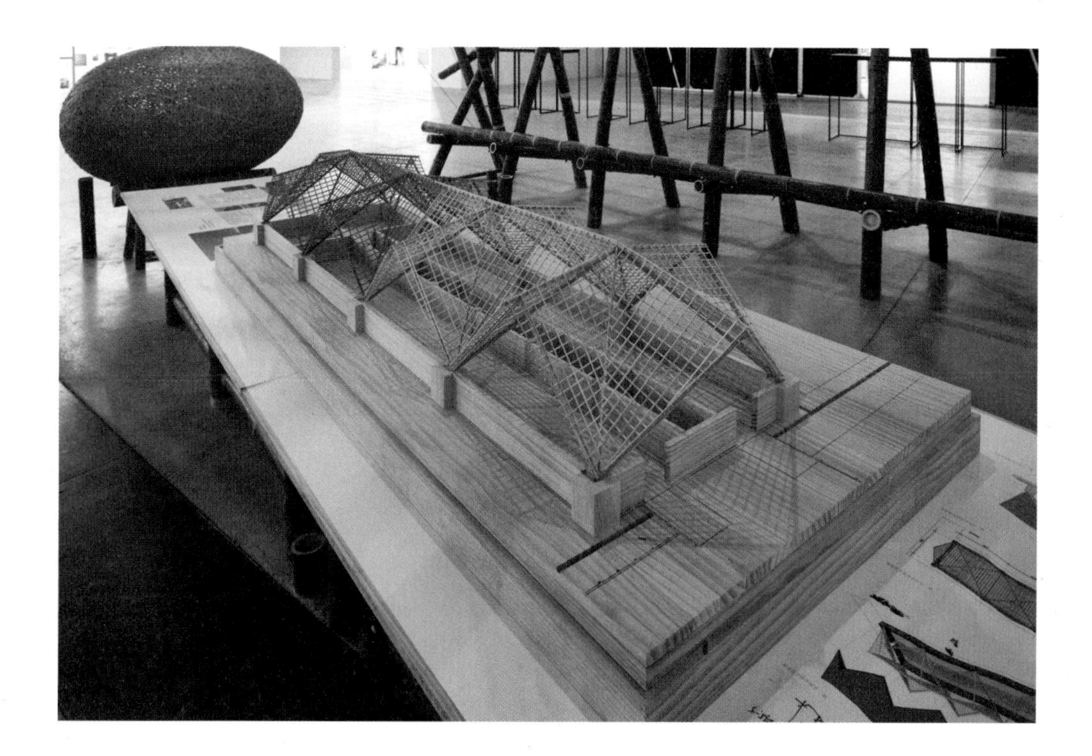

展览意在再现乡野中人在竹林的场景。作品从原竹的自然特性中展现竹的特点。展品和参观者在构筑的竹林中，回顾在地的乡村生活，以及对未来的展望。作品由临安太阳公社的村民社员合力搭建，材料和做法延自太阳公社。展览展示了公社的社区建造和自然建造的新传统，并展示六条乡建宣言。

**The exhibition aims at illustrating the picture in which villagers wander in the bamboo forests. The work expresses the characteristics of bamboo by digging its intrinsic qualities. Visitors enter the constructed "forest",look back at the local rural lives and look forward to the future. The work was built by villagers of Lin'an Sun Community, inheriting the material and building methods from the Sun Community. The exhibition demonstrated the new traditions in the artificial and natural construction of the community, and the six declarations of rural constructions.**

**陈浩如 | Chen Haoru**

杭州人，生于 1970 年代，山上建筑工作室主持建筑师。Born in the 1970s in Hangzhou, Chen Haoru is currently the principal of Architecture on Hills atelier.

# C-07

## 许村：艺术乡建的中国现场
## XU VILLAGE: SITE OF ARTISTIC RURAL CONSTRUCTION IN CHINA

“许村：艺术乡建的中国现场”部分，为艺术家渠岩的乡建团队八年来的工作汇集。展览通过文字和图片、规划与模型、录像和视频、实物与产品等视觉呈现手段，向观众呈现了“许村计划与实践”的成果，包括“许村乡建”的理论与观念、规划与设计、改造与修复、艺术与介入、民艺与抢救、启蒙与影响、教育与助学、产品与包装等等。

**“Xu Village: Site of Artistic Rural Construction in China” is a collection of the works done by Qu Yan and his team for eight years. Through texts and images, planning and models, videos and video products and other visual presentation means, the success of “Xu village plan and practice,”including the theory and concepts of “Rural Construction in Xu Village “, planning and design, renovation and restoration, art and intervention, folk art and preservations, enlightenment and influence, education and aiding-education, products and packaging and so on, is demonstrated to the audience.**

**渠岩 | Qu Yan**

作为中国当代最具代表性的前卫艺术家之一，渠岩的创作包括绘画、装置、摄影、新媒体以及艺术介入社会实践项目，参加了许多国际和国内重要的当代艺术展览。Qu Yan is one of the most representative contemporary Chinese avant-garde artists. His works include paintings, installations, photography, new media and art involved in social practice project.he participated in many important international and domestic contemporary art exhibitions.

# C-08 谢英俊 × YOU 营造及设计工作坊 HSIEH YING-CHUN × YOU DESIGN STUDIO

当建筑碰上互联网时代，建筑产业从市场需求、设计到建造将会有一个什么样的未来？谢英俊建筑师提出他的想象，将其发展累积 15 年之多元、开放、弹性之强化轻钢系统，通过使用者、设计师、建造者、销售者等各方角色参与，应用在互联网平台上，创造出一个全新的产业模式。这次展览通过举办工作坊，邀请各方建筑设计师，一同了解此系统背后的强大优势，以及如何实地操作并应用于设计，深入探讨如何看待现有建筑体系，与如何成为一个未来的尺寸定义者。

**In the internet era, what will happen to architecture industry in respect of demand, design, and construction in the future?**

**Hsieh Ying-Chun has put forward his proposal for creating a new industrial model, using the internet as a platform to all participators, such as designers, builders and suppliers, making full use of reinforced light gauge steel system, and combining high standards of properties in multiplicity, openness and flexibility. During the workshop in West Bund, we invited architects to join us, to work together and to learn the great advantages of this system by practical designing, and thus come to a new opinion towards the current building system, and towards the way of being a dimension definer in the future.**

**谢英俊 | Hsieh Ying-Chun**

谢英俊长年致力于轻钢结构房屋研发与建设工作，先后完成台湾灾后重建 1000 余户、四川 5·12 地震 500 余户房屋建设。2004 年入选联合国最佳人居环境案例；2011 年获美国柯里·史东设计首奖。 Hsieh Ying-Chun has been dedicated himself to the development and construction works of light steel structure system. Hsieh and his team has completed more than 1500 post-quake reconstruction housing in both Taiwan and Sichuan. He has been awarded Curry Stone design prize in 2011 and nominated finalist of UN-habitat's Best Practices in 2004.

# 第三章

## 南京大學

## 身份，地方性與自主更新：徽州民居的更新設計研究

### 关麓小筑

在徽州，近年来有许多知识分子，艺术家，民间人士购买老宅进行改造，用作客栈或自宅。不同于博物馆式的保护，老宅被改造的结果，不仅是物质的新生，更是以各自的方式从情感，功能上一点点地修复着乡村。这些不具备专业背景的建筑尝试，具有原生的力量，与多重意义的地方性有着绵密的关系，模糊不清的界限。这一种民间建筑实践，无论从建筑学本身，还是从理解当代乡村的角度，都是极具意义的。

我们希望作为这些事件的观察者与呈现者，将这些不应被埋没的鲜活案例反馈到学界与社会，为更多人所看见与借鉴。

我们选择徽州地区若干个有价值的改造实例，实地走访。现已经历呈坎首善宫邸与关麓小筑两站。从物质改造入手，测绘整理改造前后的基础性资料，记录改造的过程。同时进行深入观察与访谈，试图超越物质层面的讨论，探索身份和地方性与改造活动的微妙关联。

每一个个例都是独特的。关于关麓小筑，我们关注到的，是建筑的媒介性与自主生长性。最初它承载的愿景大于它本身，建筑改造成为这些愿景的媒介，反而退居次要。然而在被使用中建筑开始生长，在与人日常性的共生里获得自身的序列，超越预设。以左靖先生自宅兼碧山计划工作站为开端，经历了一次功能转型，演变为民宿，还在过渡期就具有了客人赖着不想走的奇妙魅力。小筑也逐渐在成为平静小村中一个活力的辐射点。而这些背后就是我们要探寻的东西。

团队简介：王瑶，鲁晴，王秋锐，罗晓东。南京大学，建筑学二年级学生。

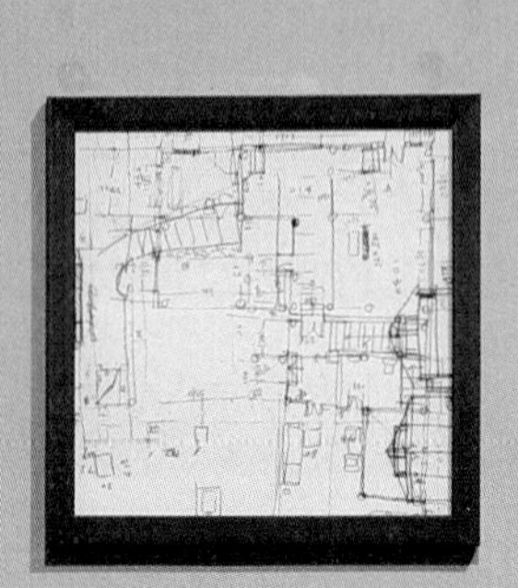

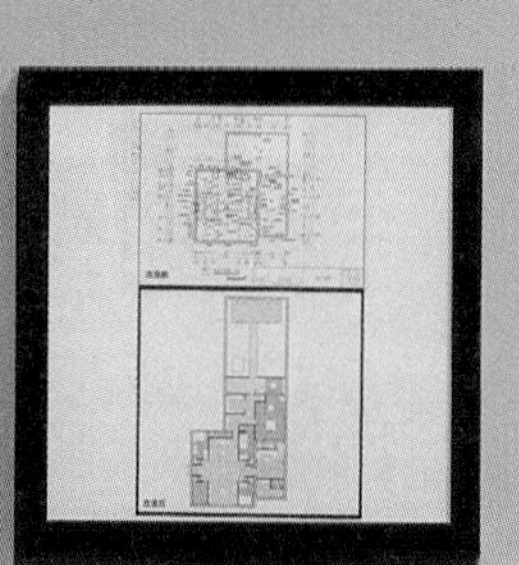

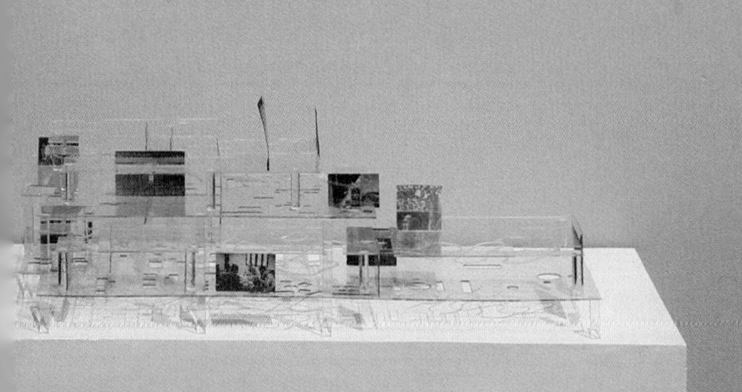

碧山書局

新碧山村的空间政治在 1949 年后开始奠定。地方自治彻底瓦解，土地所有权问题一次性解决。这些根本性的变化一直延续至今。2005 年，两位诗人到达碧山村，开始改造一幢古民居；2011 年，碧山计划落地碧山村，新碧山村的空间发生了前所未有的变化。

2005 年之后碧山村发生的空间生产行为完全是“异质化”和去“中心性”的，带有一种隐匿的回归企图。甚至，它看起来是一次对于前一场根本性变化的反抗。这些空间生产自身也处于一种博弈状态中，既生动多元，又充满悖论。在某种程度上，它形成的空间场隐隐动摇了原先的空间政治。

这次展览选取了四个空间生产案例，通过它们，我们可以看到这些空间的前生产、生产和再生产都体现了受制于政治、经济和文化上的振荡性，并在时间的参与下，继续发酵。

The spatial politics in the new Bishan Village has laid its foundation since 1949, when local autonomy completely collapsed and land ownership issues were totally resolved once and for all. Such radical social transformation continues to influence the village today. In 2005, two poets came to the Bishan Village to renovate an ancient private house. In 2011, the Bishan Project began, bringing about unprecedented spatial changes to the new village.

In my view, "heterogenization" and "decentralization" have been the dominant features of the spatial production after 2005, indicating the hidden attempt for reviving the original state. It's even more like a resistance to the radical changes in the past. These spatial productions themselves were diversified while paradoxical. To a certain extent, it has some what shaken up the original spatial politics.

From the four cases presented in this exhibition, we could see that pre-productions, productions, and re-productions of these spaces were bounded by the movements in politics, economy and culture, and will continue to evolve over time.

**左靖 | Zuo Jing**

1970 年 11 月生。策展人、出版人、乡村建设者。碧山计划的联合发起人；《碧山》杂志书主编。 Born in November, 1970. Curator, publisher, and rural construction practitioner. Co-Founder of the Bishan Project. Editor-in-Chief of *Bi Shan*.

# C-10 新木构 NEW TIMBER STRUCTURE

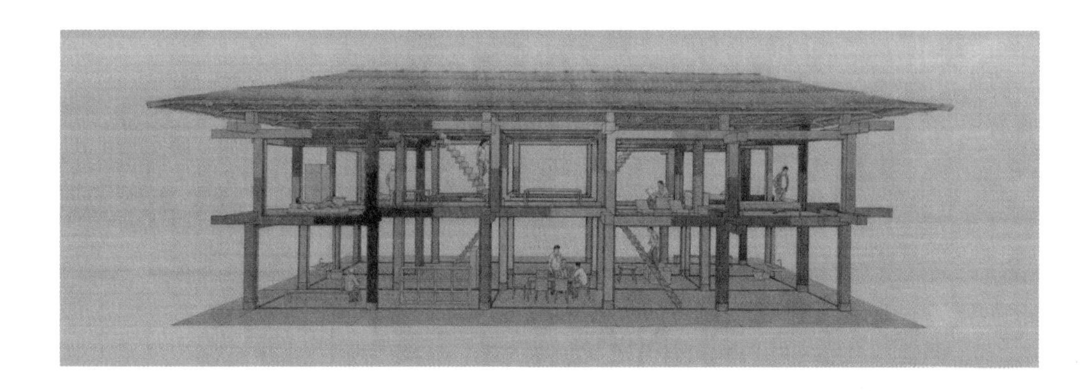

在世界曾经出现过七大独立的建筑体系中，仅中国建筑体系是以木结构为本位的。中国在很早的时候就形成了木结构建筑“标准化”的概念。这种“类工业化”的建筑方法对汇集工匠经验、加快施工进度、节省建筑成本有显著作用。21 世纪的今天，我们是否能够重新反思中国营造的思想与表征，通过创新地运用木材这种具有生命感的建材，缓和一些建筑带来的时代病症？

**Chinese architecture system is the only timber structure system among the seven independent architecture systems in the world.The timber structure modulus system appeared early in China. This 'industrialization-like' building method helped significantly in gathering craftsmen experience, accelerating construction process, and saving budget. In the 21st century, is it possible to revive the soul of traditional Chinese constructing techniques through using timber innovatively, and solve some problems brought about by modern architecture?**

**润 · 建筑工作室 | Rùn Atelier**

润 · 建筑工作室主张设计从日常美学与独立思考开始，力行“人宅互养”的建筑理念，反观传统人文，取长古今工艺，尊崇自然共生法则，悉心营造文质并美、返璞归真的当代“润”生活美学。 Based on daily aesthetics, independent thinking and architectural philosophy of mutual benefit of resident and residence, Rùn Atelier learns from both ancient and modern art and respects the natural law of coexistence. Rùn aesthetics is created carefully to achieve the beauty of subject and style, to return to purity and simplicity.

# 前瞻

## 新兴城市范式

## PROSPECTIVE

## 前瞻：新兴城市范式

“前瞻”板块展示了数字媒体文化对建筑与都市想象的催生与影响。媒体不仅记录展现城市与建筑的实践状况，推动理论、评论和学科的发展，更直接参与社会生产方式的变革。该板块旨在展示信息化时代下，新传媒、计算机参数化设计、互联网技术等对新的都市生活范式和城市公共空间的探索与引领。

## PROSPECTIVE: EMERGING URBAN PARADIGMS

This section shows the stimulation and impact of digital media culture on the architectural and urban imagination. Media not only records and presents urban and architectural practice, but also propels the development of theory, criticism, and the discipline itself, even participating directly in the transformation of the mode of social production. This section aims to explore the frontiers of new media, parametric design, and Internet technologies, and how they influence new urban living paradigms and public spaces.

沟通规划
Creative City
土地再整理
紧凑城市网络社会
Placemaking
景观生态学
城市复兴
Community Activism
创意集群
城市触媒
Adaptive Reuse
技术城市
Land Readjustment
包容性区划
景观都市主义
Adaptive Reuse
Urban Catalyst
Compact City
Creative City
都市农业
保护性再利用
社区行动主义
New Urbanism
环境导向
城市更新
精明增长
Smart Growth
Urban Renaissance

# D-01 生态都市主义装置 ECOLOGICAL URBANISM INSTALLATION

当代城市发展同时被经济发展的需要以及环境可持续的要求两方面制约。生态都市主义对这种两难挑战进行回应，提出城市化进程中需要新的道德和美学标准，以让我们的社会环境更加包容，同时对生态环境变化更加敏感。生态都市主义从哲学家瓜塔里的"生态知学"概念中找到线索，扩大了全球化背景下城市化理论的思考范围，对环境、社会公平和人类感知三个大范围内的话题进行整合。由莫森·穆斯塔法维和加雷斯·多尔蒂编辑，Lars Müller 出版社于 2010 年 5 月出版的《生态都市主义》一书是对这一范围内的新兴理论、创新设计、社会活动和前瞻性法规提案的框架系统。

**Contemporary urban development is simultaneously challenged by our needs for growth and the necessity of sustainability. Ecological urbanism confronts and responds to such challenges by calling for new ethics and aesthetics in an urbanism that is more socially inclusive and ecologically sensitive. Drawing from Guattari's ethico-political concept of ecosophy, ecological urbanism integrates a broader range of concerns around the global practice of urban development, summarized by three aspects: environment, social equity and human subjectivity. *Ecological Urbanism*, edited by Mohsen Mostafavi with Gareth Doherty and published in May 2010 by Lars Müller Publishers, is a framework for emerging theories, innovative design projects, inclusive activist practices and forward-looking regulation proposals.**

**莫森·莫斯塔法维、加雷斯·多尔蒂 | Mohsen Mostafavi, Gareth Doherty**

建筑师和教育家莫森·莫斯塔法维现任哈佛大学设计学院院长、亚历山大和维多利亚维礼设计教授。他的近作包括《生态都市主义》（已译作中文、葡萄牙语和西班牙语出版）、《城市的生命中》及《尼古拉斯·霍克斯莫：伦敦教堂》。

加雷斯·多尔蒂是哈佛大学设计学院景观建筑学助理教授及高级研究员。他的研究关注设计方法及理论，着重于人类学、地理学和设计的融合。多尔蒂和莫斯塔法维共同编著了《生态都市主义》，他也是《新地理学》杂志的创始编辑和该刊第三期《色彩都市主义》的主编。

An architect and educator, Mohsen Mostafavi is the Dean of the Harvard University Graduate School of Design and the Alexander and Victoria Wiley Professor of Design. His recent publications include *Ecological Urbanism* (translated into Chinese, Portuguese, and Spanish), *In the Life of Cities*, and *Nicholas Hawksmoor: The London Churches*.

Gareth Doherty is an Assistant Professor of Landscape Architecture and Senior Research Associate at Harvard University Graduate School of Design. Dr Doherty's research and teaching focus on design methods and theories, with an emphasis on the synergies among anthropology, geography, and design. Doherty edited *Ecological Urbanism* with Mohsen Mostafavi, and is a founding editor of the *New Geographies journal* and editor-in-chief of *New Geographies 3: Urbanisms of Color*.

SSTA—ESPSNAI
ISELOASLCEHGI
SLLSMLA—AH

# D-02

## 2015 上海嗅觉地图——嗅觉游戏和档案
## SMELL GAME & ARCHIVE

这个互动装置的初衷在于让观众意识到可以通过嗅觉来理解上海这座城市。游戏参与者闻到了西岸艺术中心周围各处采集的气息，并被邀请对这些气味做出评价。这个展览试图让人们体验到城市生活中一种常常被忽视的潜在信息——气味。

**The purpose of this interactive installation is to enlighten the visitor that Shanghai can be understood by using one's nose. The participants smelt smells collected from diverse sites surrounding the West Bund Art Centre and were invited to comment on them. The exhibit makes people more conscious of an additional layer of information in urban living - the smells.**

**西塞尔 · 图拉斯 | Sissel Tolaas**

挪威裔艺术家西塞尔 · 图拉斯是一位职业多面手，她有化学和艺术两方面的专业背景，从 90 年代起专攻气味。她从 2004 年起，在柏林的气味研究室专事用气味促进交流。她以实验室作为对话平台与其他各领域专家都有合作，力图转变人们对气味的认知和品尝气味的方式。图拉斯弥合了气味领域和科学间的缝隙，将两者都带入我们生活的现实。 Tolaas has background in chemistry & art. She has been 100% focused on the smell since the beginning of the 90s. She run the SMELL RE_searchLab Berlin, on smell & communication since 2004.The Lab is a "station of dialogues," in other words a place to establish a continuous dialogue between a big range of other experts. The goal of the Lab is to change the existing approach to "our noses & smells and the process of smelling." Tolaas bridges the gap between the smell industry and science. She brings them both in touch with the reality we live in.

水城慢生活
Slowater City

# D-03 东京二零五零 TOKYO 2050

TOKYO2050 // Nomadic Life on Energy Geography
2050
2011

1. 水城慢生活

东京曾经是可以与威尼斯媲美的“水上都市”。特别是隅田川东边的江东墨田地区，孕育出了与水密切关联与共生的独特城市风景和文化主题。本提案把依旧继承着原有的水域环境（河流、沟渠、运河、海）和其历史重叠在一起，同时不断发掘出多样化的生活文化资产，提出了面向未来、具有创造性的地区设计方案。

2. 能源几何的游牧生活

在不久的将来，2050年，以东京作为主线，从如何有效利用能量与热量来展开观点，重塑都市构造及生活形态来进行提案。以有着小型加工场和住宅交错在一起的东京湾沿岸部的大田区蒲田周边为对象地块，如何活用地形和生产运动所产生的“能源、几何”来反映都市生活形态，并且对这些起了引导作用的都市形态部分做了尝试性的表述。

3. 向寿町和上海学习

在寿町有着让生活变得丰富多彩的实践活动，我们认为这些都是对将来的城市建设有利的。作为实践活动具体的案例包括有效利用道路空间、单身人士的相互扶助活动之类。

1. Slowater City. Tokyo was once a city like Venice, a "Water City". The area east of the Sumida River, Kōtō-Sumida, was the birthplace of a unique urban scenery and cultural topos closely tied to the water. Due to modernization, this area transformed and lost much of its original personality. Our proposal seeks to revisit its legacy as an environment of water, history and culture, and proposes methods for creative, future-oriented development of this area.

2. Nomadic Life on Energy Geometry. This is a proposal to remodel the urban structure and form new lifestyle of Tokyo in 2050, near future, based on efficient energy and heat usage. Site is Kamata area, Ota Ward in a bay side area of Tokyo, where small factories and houses are mixed together in high density. We tried to create an image of urban life on "energy geometry", which is formed by geography and production activity, and also to create a townscape, which leads to this urban life.

3. Learning from KOTOBUKI and SHANGHAI. We were able to find attractive practices that could develop into future town planning methods for enriching our life. For example, streets are utilized effectively and mutual aid is seen between single living bachelors.

**阵内秀信研究室（法政大学）+ 山口尚之（+・建筑设计事务所）、首都大学東京小泉雅生研究室 + 門脇耕三 + 猪熊純、神奈川大学曾我部昌志・吉冈宽之研究室 + 西岸双年展研究团队 | Hidenobu JINNAI Lab. (Hosei University) + Naoyuki YAMAGUCHI (TASTEN Architects), Masao KOIZUMI Lab. Tokyo Metropolitan University, Kozo KADOWAKI, Jun INOKUMA, Masashi SOGABE and Hiroyuki YOSHIOKA Lab. (Kanagawa University) + Project team for the West Bund Biennale**

2020
SHANGHAI
2040
SHANG HAI
上海2040

# D-04 上海二零四零 SHANGHAI 2040

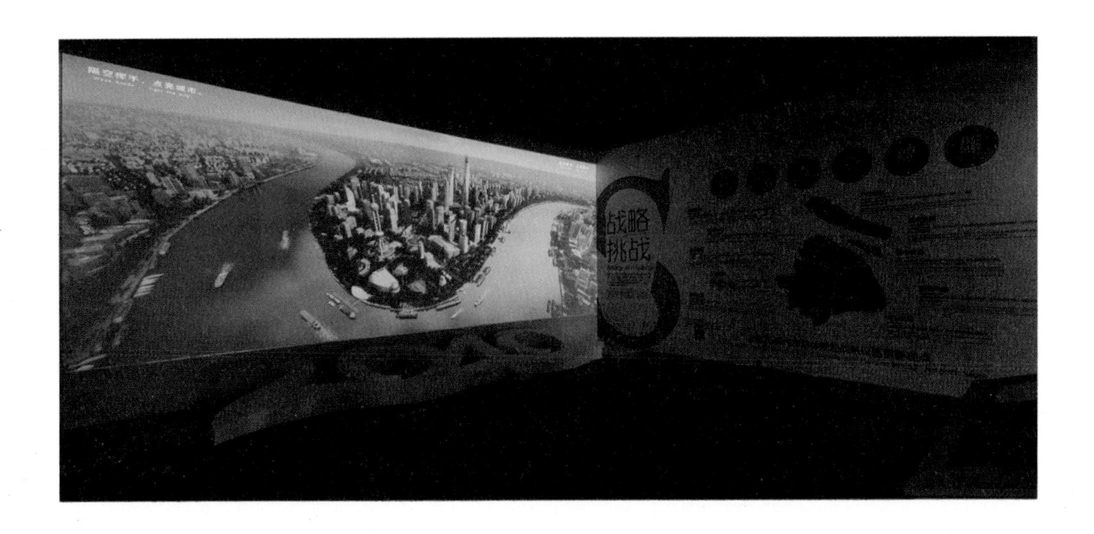


2040 年，恰逢上海开埠 200 年、浦东开发开放 50 年，" 领中国开放之先 " 的上海，往哪里去？以开放谋求全球竞合力。开放是城市发展的胸襟和格局，开放是上海的传统和优势，是上海融入全球化网络的必然。以绿色提升持续发展力。绿色是城市可持续发展的必由之路，当我们沐浴了工业化的雾霾之后，更加珍惜每天所享有的大气、水和土壤的安全、洁净。以关怀塑造城市独特魅力。市民幸福是城市发展的本质追求，人文关怀是城市魅力的本源所在。多元包容是城市文化的象征，更是城市文明的内涵。 2040 年的上海，将成为一个令更多人向往的、更加绿色、更加开放、更加美好的全球城市。

The year of 2040 marks the 200th anniversary of Shanghai-Port Opening to the world and 50th anniversary of Pudong-District Development Announcement. Then what is the future of this pioneering city? Opening-up brings us a global competitiveness. The open-minded way is not only the best showcase of Shanghai when comes to its ambitious development layout but also its distinctive tradition and advantage as well as one of the results integrating into globalization. Green-energy brings us a sustainable power source for development. A city cannot achieve a sustainable development goal without using green energy. What we have learnt from suffering the industrial fog is the changes on our altitude which gives more importance to the safety and cleanness of air, water and soil. Humanistic-care brings us a unique city character. Citizen welfare is the ultimate goal for city development. Diversity symbolize a city culture and its connotation. Shanghai, in the year of 2040, is believed to be a city attracting more people with its more open-minded, green and beautiful future.

**上海市城市规划设计研究院、上海同济城市规划设计研究院 | Shanghai Urban Planning & Design Research Institute, Shanghai Tongji Urban Planning & Design Institute**

ON BROADWAY
The interactive installation ON BROADWAY represents digital traces of life in the 21st century city through a compilation of images and other data collected along the 13 miles of Broadway that span Manhattan. The result is a new type of street view, created from the activities of hundreds of thousands of people.
http://on-broadway.nyc
THE CLOISTERS
COLUMBIA UNIVERSITY
TIMES SQUARE
HERALD SQUARE
MADISON SQUARE PARK
UNION SQUARE
CITY HALL

# D-05 百老汇街 ON BROADWAY

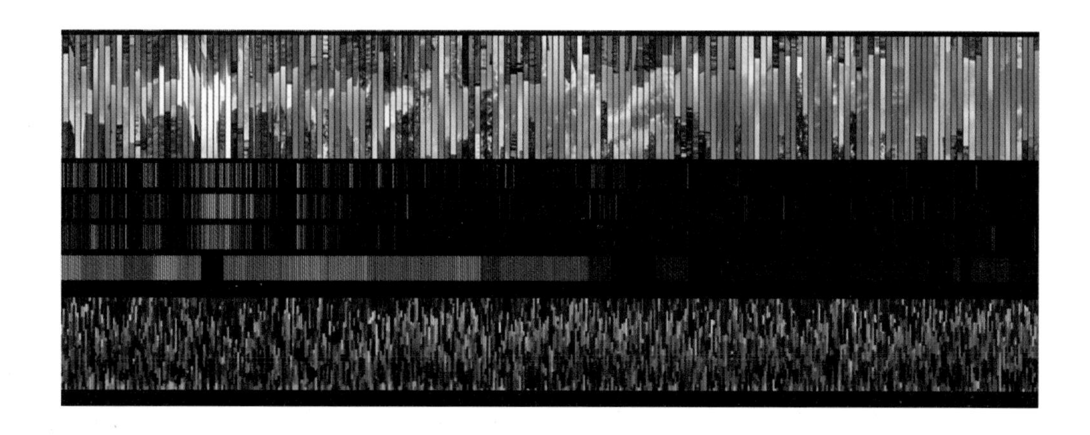

我们在此研究项目中只关注纽约的百老汇街。如同人体脊椎，百老汇街穿越了曼哈顿岛。我们在研究中纳入了街旁的区域，从而可以捕捉到周边的活动。为了限定这片区域，我们沿着百老汇街中心每隔 30 米布点，并以该点为中心做一个 100 米宽的切片，结果就是一个 21 390 米长、100 米宽的脊椎形态。我们用这个形态坐标来过滤取自整座纽约城的数据。我们在研究过程中使用了大数据，比如来自 Instagram 平台的 66 万张分享自百老汇街沿途的照片、2200 万次出租车的上下客数据等等。总体上看，这项研究使用了超过 4000 万张图片和数据点来再现了一条街道。

**In our project we focus on a single street in NYC: Broadway. Like a spine in a human body, Broadway runs through the middle of Manhattan Island curving along its way. We wanted to include a slightly wider area than the street itself so we can capture also the activities nearby. To define this area, we selected points at 30 meter intervals going through the center of Broadway, and made 100 meter wide slices centered on every point. The result is a spine-like shape that is 21,390 m long and 100 m wide. We used the coordinates of this shape to filter the data we obtained for all of NYC. Image and data include 660,000 Instagram photos shared along Broadway during six months in 2014, Twitter posts with images, Foursquare check-ins since 2009, Google Street View images, 22 million taxi pickups and drop-offs in 2013, and economic indicators from US Census Bureau (2013). Overall, the project uses over 40 million images and data points to represent a single street.**

**达尼尔 · 戈德迈耶、莫里兹 · 史蒂芬尼、多米尼库斯 · 鲍尔、列夫 · 马诺维奇 | Daniel Goddemeyer, Moritz Stefaner, Dominikus Baur, Lev Manovich**

戈德迈耶与鲍尔研究互动媒介与体验式设计；史蒂芬尼专长于信息可视化；马诺维奇是创新软件研究机构的负责人，同时也是纽约城市大学研究生院中心的教授。 Goddemeyer and Baur are designers working on interactive media and experience design projects. Stefaner is an expert in information visualization. Manovich is a Director of Software Studies Initiative and Professor at The Graduate Center, City University of New York (CUNY).

Thermodynamic Madrid. A new g
热动力马德里·美好新生活
An experimental alternative to the policies applied to the renovation of histo
针对城市历史中心区翻新改造的试验性策略

# D-06 热动力马德里·美好新生活 THERMODYNAMIC MADRID: A NEW GOOD LIFE

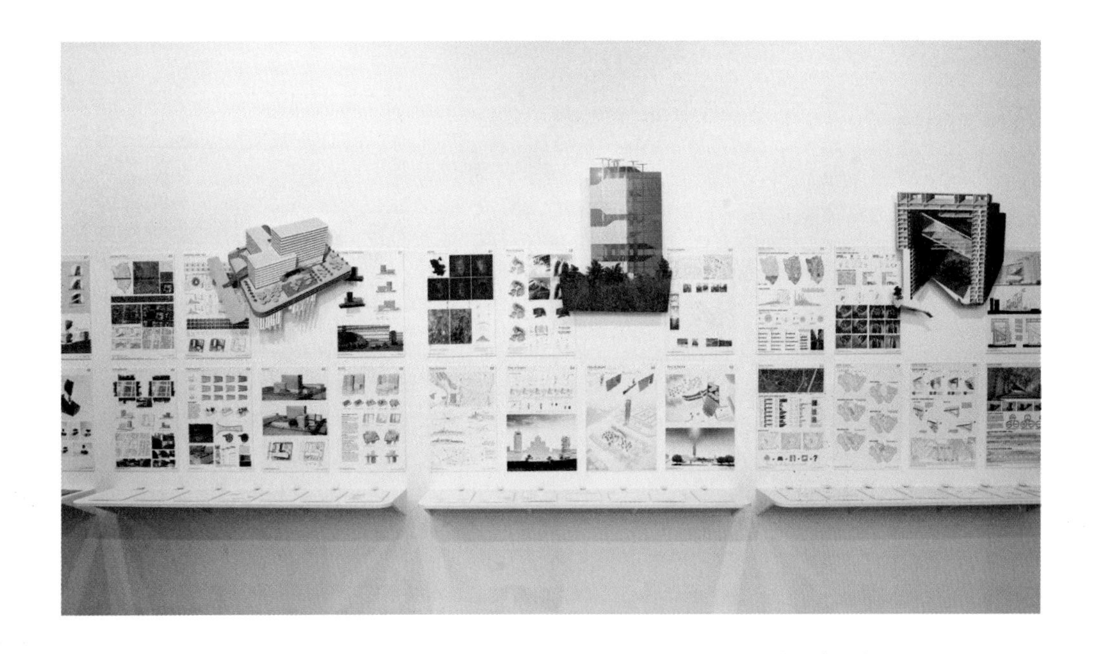

“热动力马德里·美好新生活”展位，旨在从方法和尺度层面来探求一种既不同于保护主义那样缺乏建设性的极端保护，也不同于暴力式文化铲除的历史中心区改造策略。该设计着眼于可同时应用在建筑和城市尺度的参数化技术手段，尤其是针对辐射扭矩和通风环境的分析，从而定义出一种不必更改现存密度和公共空间配置，也不伤害历史遗迹的公、私空间品质优化方案。

**The exhibition "Thermodynamic Madrid: A New Good Life" aims to explore new methods and scales to address the renovation of historic centers outside both the radical preservationism of the protectionist theories as destructive practices and social expulsion of sventramento. The studio aims to test parametric techniques applied simultaneously to urban space and building space (essentially analyzing radiation torque / ventilation) to identify quality improvement strategies applied to the public and private realms without modifying the existing density or the configuration of the public spaces (and maintaining the historic patrimony).**

**伊纳吉·阿巴罗斯 | Iñaki Ábalos**

哈佛大学设计研究院教授。Professor, GSD, Harvard University.

New York
紐約

# D-07 能量试点工程 EMBODIED ENERGY PILOT PROJECT

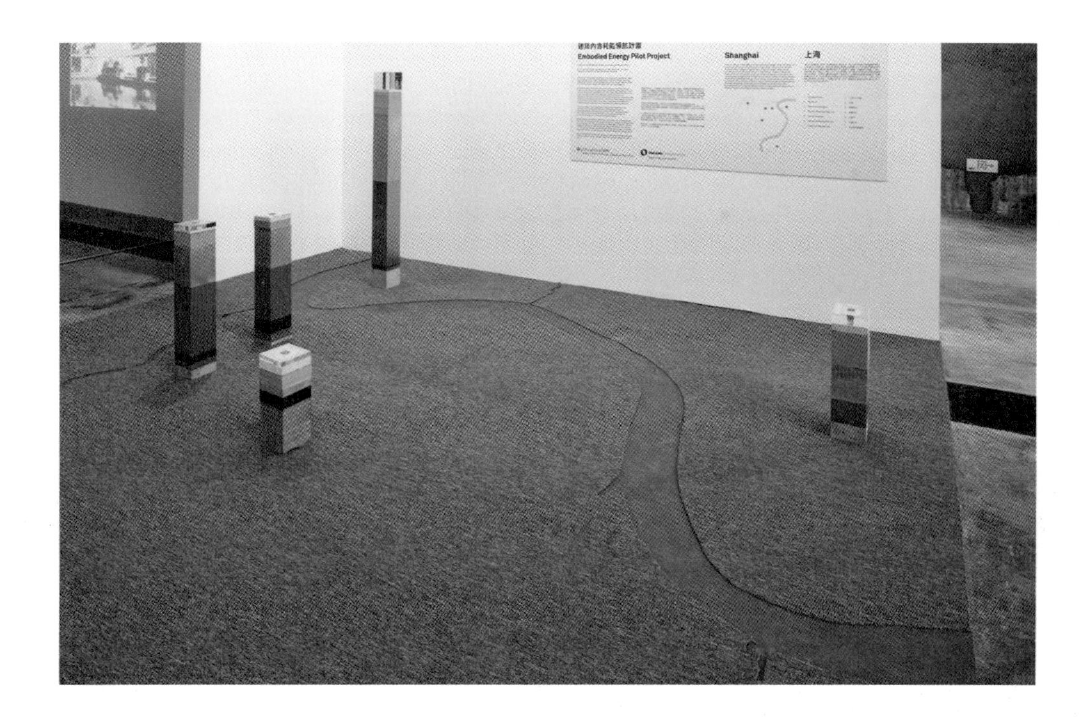


本展位对上海和纽约的建筑能量进行可视化。上海的可视化包括一张含有六座高层建筑的城市地图，它呈现了这六座代表性建筑里的能量。纽约的可视化包括两张图：曼哈顿建筑的能量图解，以及典型街道交叉口的能量图解。这些图档勾画了我们试图在从六座代表性塔楼到软木楼层图，再到打印图解这个过程中创造出来的特殊元素。

**This exhibition involves visualizations of embodied energy in Shanghai and New York. The visualization of Shanghai involves a rough map of the city with six physical towers representing the relative embodied energy in six representative buildings. The visualization of New York involves two drawings: an embodied energy map of the buildings in Manhattan, and an embodied energy drawing of a typical street intersection. This document outlines the specific elements that we would like to create, from the six physical towers to floor map made of cork to the printed drawings.**

**大卫·本杰明 + 哥伦比亚大学建筑、城规与保护研究生院 | David Benjamin + GSAPP**

GSAPP 结合了先锋学科实验，坚定地投入到对世界以及我们这个时代迫切问题的研究中去。 The Graduate School of Architecture, Planning and Preservation at Columbia University combines pioneering disciplinary experimentation with an uncompromising engagement with the world and the urgent questions of our time.

ASA IN ASIA
TERREFORM ONE

# D-08 城市工程布鲁克林 2110 URBANEERING BROOKLYN 2110

“布鲁克林 2110”项目主要采纳了一种假设，即城市所需资源都应能在其物理边界内得到。本设计展现了一种加强版的布鲁克林，重塑了其中的食物、水、空气、能源、垃圾、交通和房屋来满足市民所需，并支持各种形式的生命。设计策略包括用垂直农业和融合了住宅的基础设施替代过时的建筑；过去的街道变成了蜿蜒的、可居住的空间大动脉，并集成了可更新能源、软垫交通工具和食品生产温室。城市的平面采用了现存的街道分隔作为新网络的起点，而对这些过时街道的改造让我们能插入切实可用的生态走廊。这些手法并不是为了全面设计一个未来城市，而是要开启讨论的平台。我们认为未来需要的是可循环的能源网络和俊美建筑的结合——未来会到来，而我们的准备和所有人的贡献是未来的关键。

**Our primary assertion for Urbaneering Brooklyn 2110 is that all necessities are provided inside its accessible physical borders. We have designed an intensified version of Brooklyn that supplies all vital needs for its population. In this city, food, water, air, energy, waste, mobility, and shelter are radically restructured to support life in every form. The strategy includes the replacement of dilapidated structures with vertical agriculture and housing merged with infrastructure. Former streets become snaking arteries of livable spaces embedded with renewable energy sources, soft cushion based vehicles for mobility, and productive green rooms. The plan uses the former street grid as the foundation for new networks. By re-engineering the obsolete streets, we can install radically robust and ecologically active pathways. These operations are not just about a comprehensive model of tomorrow's city, but an initial platform for discourse. We think the future will necessitate marvelous dwellings coupled with a massive cyclical resource net. The future will happen, how we get there is dependent upon our planned preparation and egalitarian feedback.**

**Terreform ONE**

Terreform ONE 是一家倡导城市中智能设计的非盈利设计机构，其目标是通过创意项目和公众教育为纽约及世界其他城市提供思考新城市环境的可能性。队伍包括科学家、艺术家、建筑师、学生和其他各领域人士，以探索从社会到生态的设计框架。 Terreform ONE is a non-profit design group that promotes smart design in cities. Through our creative projects and outreach efforts, we aim to illuminate the environmental possibilities of New York City and inspire solutions in areas like it around the world. We are a unique laboratory for scientists, artists, architects, students, and individuals of many backgrounds to explore and advance the larger framework of socio-ecological design. The group develops innovative solutions and technologies for local sustainability in energy, transportation, infrastructure, buildings, waste treatment, food, and water.

建设规模：占地10万㎡
总建筑面积24万㎡
开工时间：2011年
竣工时间：2017年
设计单位：非常建筑（塔前广场及文化地块）
华东院（商办地块）
龙华地区综合改造
作为徐汇滨江地区的重要组成部分，以轨道交通11、12号线建设为契机，围绕千年古刹龙华寺（公元247年）和龙华塔保护利用，全面推进涵盖土地前期开发、地下空间工程、市政道路和桥梁工程、防汛墙改造及公共绿化的龙华地区综合改造工程，进一步改善城市面貌，实现地区产业功能升级。
塔前广场及文化项目将强化龙华塔的核心作
振兴传统文化。
一云锦路跑道公园

Chris Hardie
占地10万㎡
总建筑面积24万㎡
开工时间：2011年
竣工时间：2017年
设计单位：
华东院（商办地块）
PM2.5

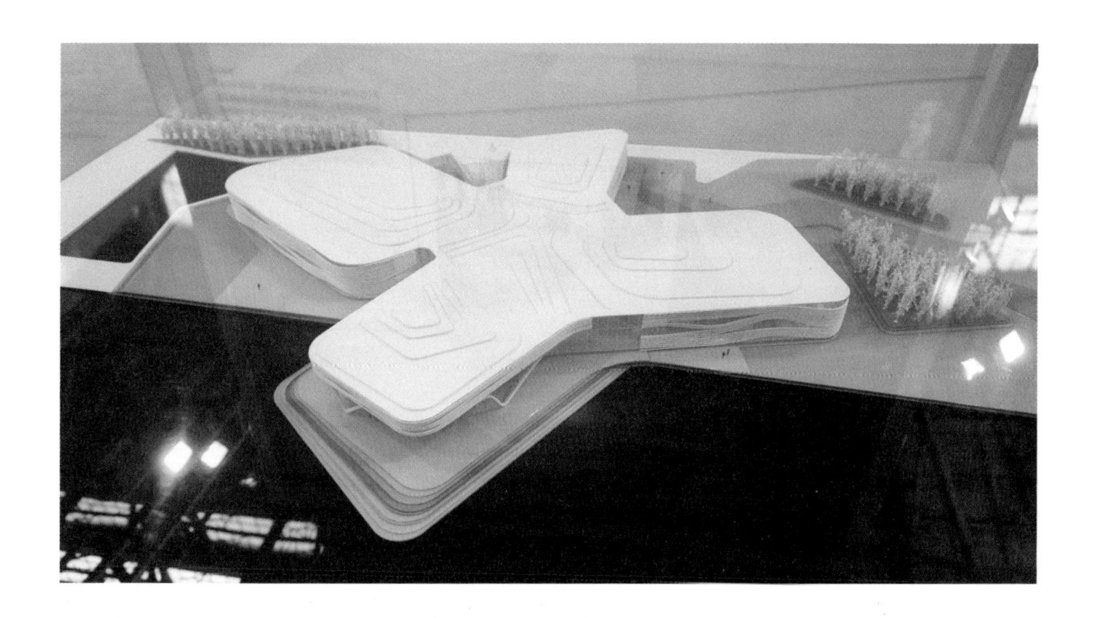

展出作品为上海西岸在工业遗存保护利用、滨江公共空间和城市功能组团建设方面具有突出代表性的实践案例，包括：1. 西岸传媒港，以"梦中心"为旗舰，打造高端文化传媒业集聚区；2. 西岸万国金融中心，以创新金融为核心，打造新型金融聚集区；3. 西岸华鑫金融中心，汇集多种设施的城市综合体；4. 西岸油罐艺术公园，以公共艺术为主基调的文化中心；5. 龙美术馆·西岸馆，从工业历史遗存形态中抽取伞形元素，是国内规模最大的私人美术馆；6. 余德耀美术馆，一座虚实对比、新旧融合的国际化标准的当代艺术美术馆；7. 西岸美术馆，滨江公共开放空间的有机组成部分，打造分享艺术体验的平台；8. 水边剧场，集小型剧场、展示大厅和多功能活动室为一体的公共设施。

**All of the arts presented here are examples of renewing industry areas and construction of public space with gathered function areas: 1. Dream Center, the principal part of West Bund Media Park; 2. the core idea of WFC development, creating a brand new aggregation area of service industry, characteristic business, and entertainment recreation; 3. Huaxin Financial center; 4. Oil Tank Theater & Café, presenting public art elements and contemporary art style; 5. Long Art Museum, the biggest private museum in China; 6. Yuz Museum is a kind of contemporary art museum; the hangar used before was regarded as the main exhibition hall and designer expanded the two side of hanger with glassed, which could make an atmosphere with characteristic of abstract and specific elements; 7. West Bund Museum is constructed by a way of overlay in order to melt the museum into the environment and constructions; 8. Designers have melt the poetic elements into Waterfront Theater which shows concise North Europe culture and implicit characteristic in China and it presents like a butterfly beside the Huangpu River.**

**上海西岸开发（集团）有限公司 | Shanghai West Bund Development Group CO., LTD**

"西岸"是徐汇滨江地区为建设世界级滨水新城区所打造的地区品牌。上海西岸开发（集团）有限公司，是全面负责并实施徐汇滨江地区综合开发建设的国有企业集团。 West Bund is represented by Xuhui Binjiang district which has been built in order to construct the world city beside the bank of Huangpu River. Shanghai West Bund is responsible for developing Xuhui Binjiang district.

Vertical Forest & Forest City

# D-10 垂直森林 & 森林之城 VERTICAL FOREST & FOREST CITY

垂直森林是城市内垂直致密化造林的一种新模式，旨在现代都市中，脱离城市本身的喧嚣，创造一个强大的生物多样性的系统。基本概念是在城市中心区，不通过侵占城市土地在三维的空间层面增加城市绿化率。森林之城汇集了可持续发展（反重子策略），生物多样性（不同物种的综合共居），以及共荣之心（集合社会力量）。森林之城是一座垂直城市，将上百顷的真实森林聚集在一两千平米的城市地面上。

**Vertical Forest is a model for a sustainable residential building, a project for metropolitan reforestation that contributes to the regeneration of the environment and urban biodiversity without the implication of expanding the city upon the territory. It is a model of vertical densification of nature within the city that operates in relation to policies for reforestation and naturalization of large urban and metropolitan borders. Forest City is a new city which mix sustainability (anti-baryon strategy), biodiversity (multiplication and cohabitation of living species) and empathy (bridging social capital). Forest City is a vertical city, which gathers hundreds of hectares of a real forest in one or two square kilometers of an urban surface.**

**博埃里建筑设计事务所 | Stefano Boeri Architetti**

博埃里先生是米兰理工大学城市学教授，曾执教于美国哈佛大学、哥伦比亚大学、麻省理工大学、荷兰贝拉罕建筑研究所、莫斯科斯特来卡等国际著名建筑院校。Stefano Boeri (born in 1956) is Professor of Urban Design at the Politecnico di Milano. He has taught as visiting professor at Strelka Institute, Harvard GSD, Berlage Institute and Architectural Association among others.

# D-11 天空花园 SKY GARDEN

ASA 建筑设计事务所展示了他们多年来关于现代空间与超高层建筑结合的研究。这项研究，作为柯布西耶建筑语言深入研究的一部分，致力于形态的可逆性。它定义了一系列连续几何体，它们同时确定了一个内部空间：一个光线的贮藏所，以及一个如智慧确切且精彩的游戏一般的外部空间。这种可逆的内部外部空间对 ASA 来说是城市空间的关键。ASA 长期以来已尝试设定可逆的形态，这种可逆的形态使得柯布西耶的建筑语言的表达能力回到公共的空间上，并且可以成为调和现代空间与城市的解决方法。

**ASA architecture atelier presents the research conducted over years on Modern Architectural Space and the Tower. This research, originating from an in-depth study of the Corbusian language, focuses on reversible forms. These can be described as continuous geometric forms, which simultaneously define an inside: a receptacle of light, and an outside a correct and magnificent play in an equivalent manner. This possible reversibility is for ASA an urban issue. ASA has hypothesized for a long time that reversible form, allowing the expressive Corbusian language to turn towards public space, would be a solution to reconcile modern architectural space and the city.**

**朱迪特・罗特巴赫、劳伦・所罗门 | Judith Rotbart, Laurent Salomon**

朱迪特・罗特巴赫 / 建筑师，1998 年获法国国家建筑师文凭，曾获法国国家建筑学院颁发的最佳毕业奖，自 2008 年起于法国诺曼底国立高等建筑学院教授建筑设计。

劳伦・所罗门 / 建筑师，1979 年获法国国家建筑师文凭，1982 年至 2014 年于法国巴黎美丽城国立高等建筑学院教授建筑设计及理论，并在诺曼底国立高等建筑学院继续他的教学。

Judith Rotbart /She has been a qualified architect since 1998. She received the French Academy of Architecture's prize for the best diploma. She has been teaching architectural design since 2008 at the École Nationale Supérieure d'Architecture de Normandie.

Laurent Salomon/He is an architect, qualified in 1979. He taught architectural design and design theory from 1982 to 2014 at the cole Nationale Supérieure d' Architecture de Paris-Belleville and is continuing his teaching at the École Nationale Supérieure d' Architecture de Normandie.

事半“公”倍
本次展览是对进行中的项目做一个切片的观察。
中国的城市建设逐步进入理性和反思的时代，在城市更新这个大
探索运作项目的各种可能性。建筑师在项目初期介入，帮助开发
中突破传统模式的局限性。Public Business为“公众的事务”
”，将公众利益和待开发空间交织共生变成可能，在这个过程
Public Business留下一抹空白。
我们将本次展览命名为“事半公倍”。和以往的开发项目不同的
期已经介入开发，设计的工作只完成了一半。开发商单纯追逐建
公民意识觉醒的今天已经成为过去，只有将建筑的社会性和公共
后，才能带来开发商和公众的双赢——利益和公共性价值的倍增
这次展示是在“事半”的情况下，围绕“公倍”这个话题展开的
才能实现真正意义上的“公共性效能的倍增”。
OCT华侨城
HIGHPOWER 天祥
14
15

# D-12 事半“公”倍 PUBLIC BUSINESS

本项目的挑战是，如何在 20 的高容积率下，创造一个宜人的环境和商住综合体。本案将艺术中心放在裙楼顶部，形成一个高端的生活场所，同时提供了一个上部的另一个“首层”，和一个有效地将功能垂直叠加在串联起来的中庭空间。这一策略使得高效运营的高密度综合体成为可能。

**The challenge of this project is to provide a pleasant environment for both commercial and living spaces on a very dense site, meeting the needs of these contrasting programs while fitting them together on a site with an FAR that may reach as high as 20. Our proposal is to concentrate the more commercial spaces in a large podium, which will then be capped with an art gallery level that can accommodate high-end spaces for living while also serving as a "second ground floor". Within the podium, different programs are vertically superimposed in a series of void spaces. These strategies will make it possible for a highly dense complex to function efficiently.**

**天祥 + 华侨城 | Shanghai Highpower Instrial Co., Ltd+Overseas Chinese Town Co., Ltd**

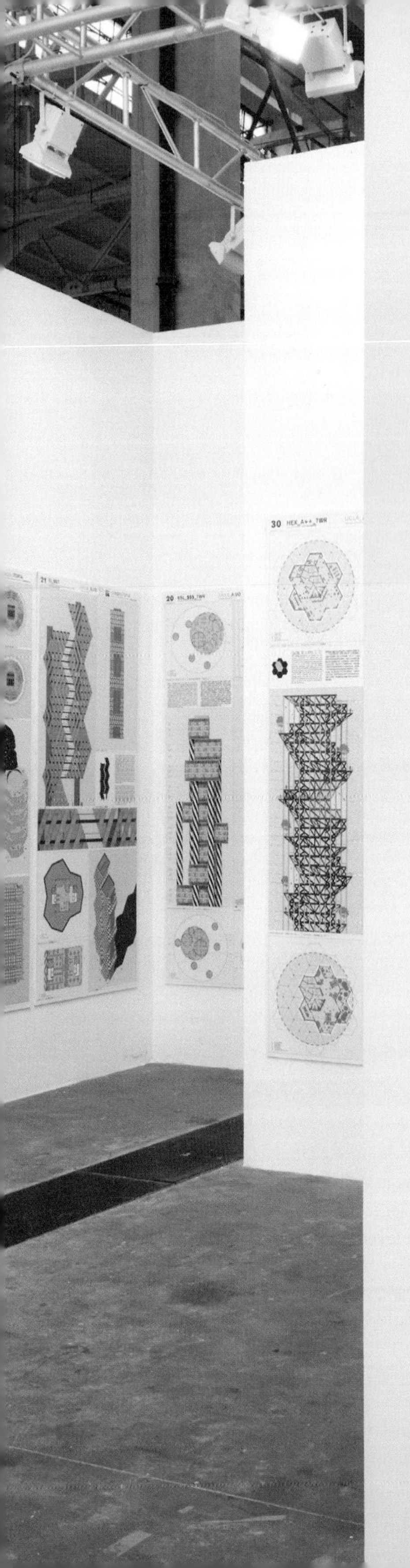

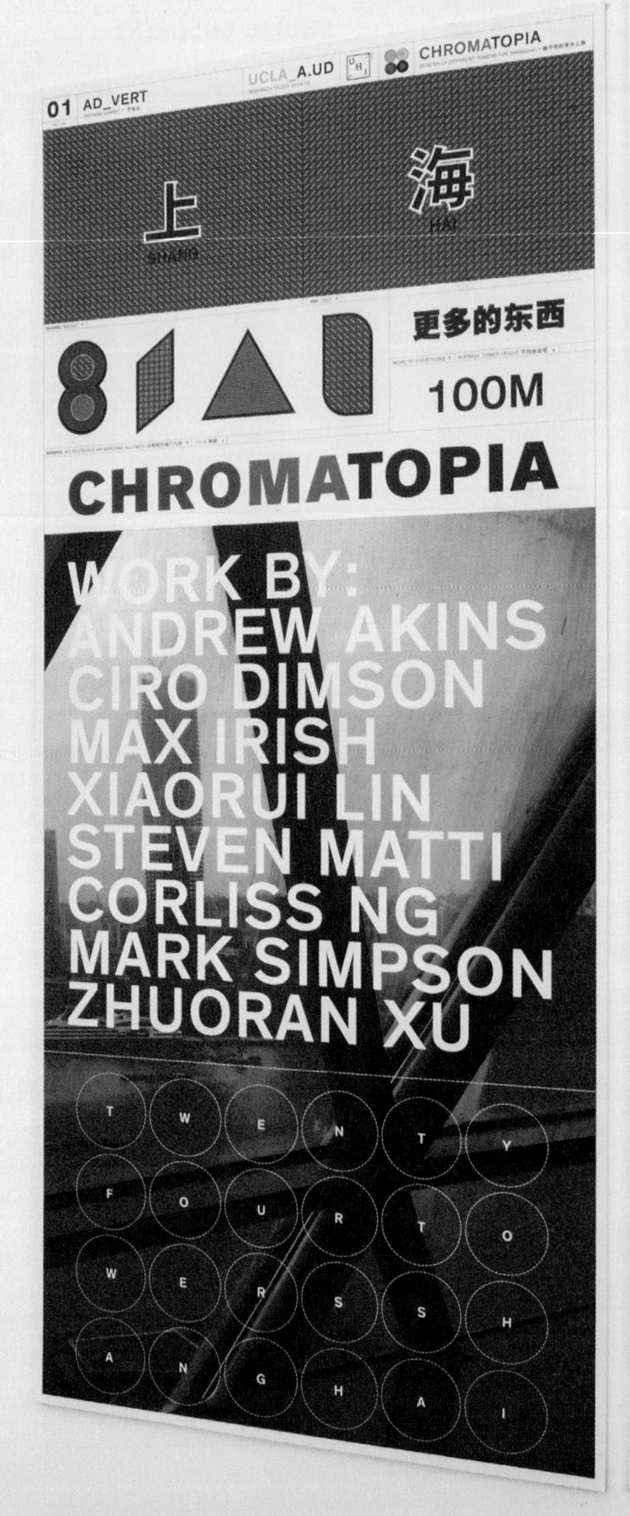

01 AD_VERT
UCLA_A.UD
CHROMATOPIA
上
SHANG
海
HAI
更多的东西
100M
CHROMATOPIA
WORK BY:
ANDREW AKINS
CIRO DIMSON
MAX IRISH
XIAORUI LIN
STEVEN MATTI
CORLISS NG
MARK SIMPSON
ZHUORAN XU
T W E N T Y
F O U R T O
W E R S S H
A N G H A I

# D-13

## 色彩乌托邦：上海的多样高楼
## CHROMATOPIA: GENERALLY DIFFERENT TOWERS FOR SHANGHAI

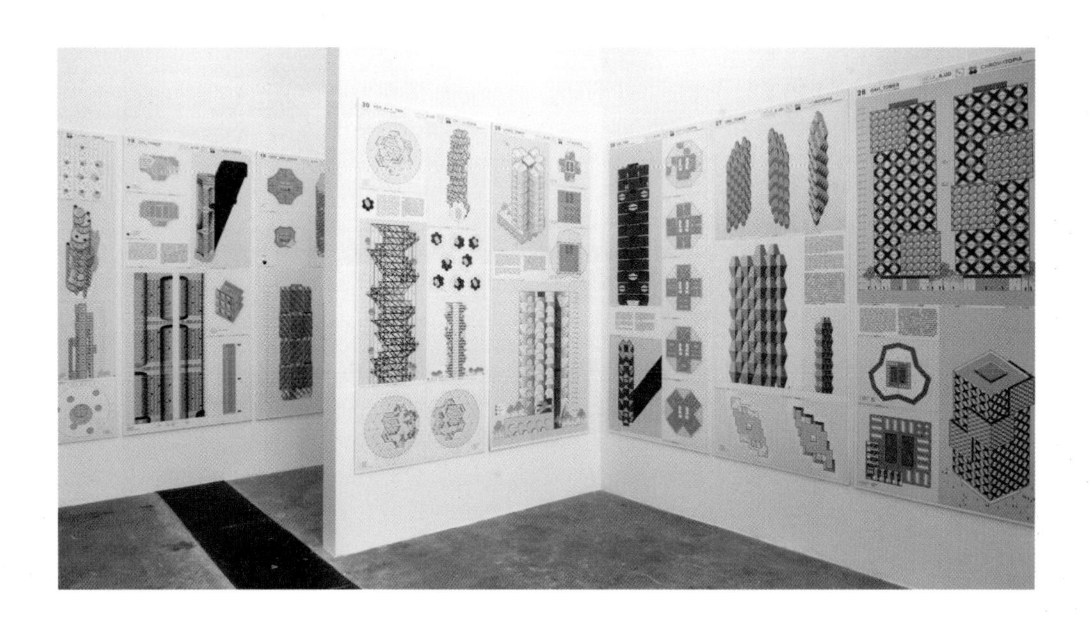

提到高楼林立的大都市，不得不说到上海。截至 2015 年，上海已拥有超过 1400 座高于 100 米的建筑。尼尔·德纳利教授带领他的加州大学洛杉矶分校建筑与城市设计学院 (UCLA A.UD) 研究工作室 2014-2015 届的学生们，在对上海的高层建筑和垂直密度长达一年的调查研究后完成了该项目。此次展出的是该团队针对上海的探索性设计（虚构设计结合经济考量）的建筑物集合。这些设计都伴随着对不同现实情境议题的描述，将全球化大都市、政治整合、社会流动性、外来劳工和无形货币等大背景融于 24 座高层建筑的设计形式中。

**What is left to be said, in 2015, about the state of the high rise, especially concerning Shanghai, a city that has more than 1400 buildings over 100 meters tall? Neil Denari and his students in the UCLA A.UD Research Studio 2014-15 investigate this question during a year-long study of high rise towers and vertical density in the city of Shanghai. The result is a collection of speculative designs (fictional and financial) and accompanying scenarios presented against a backdrop of topics including the global city, political conformity, social mobility, migrant labor, and invisible currencies — all in the form of 24 high rise towers.**

**加州大学洛杉矶分校研究工作室 2014-15 | PVG × LAX + UCLA Research Studio 2014-15 | PVG × LAX**

加州大学洛杉矶分校研究工作室 2014-15 | PVG × LAX/ 加州大学洛杉矶分校建筑与城市设计学院 (UCLA A.UD) 不断探索当代建筑学和城市规划的相关问题，尤其侧重于对先进设计的研究。 UCLA Research Studio 2014-15 | PVG × LAX/The UCLA Architecture and Urban Design Research Studio pursues issues confronting contemporary architecture and urbanism.

F.
E.
E.
L.

# D-14

## 感觉马德里
## F.E.E.L. MADRID

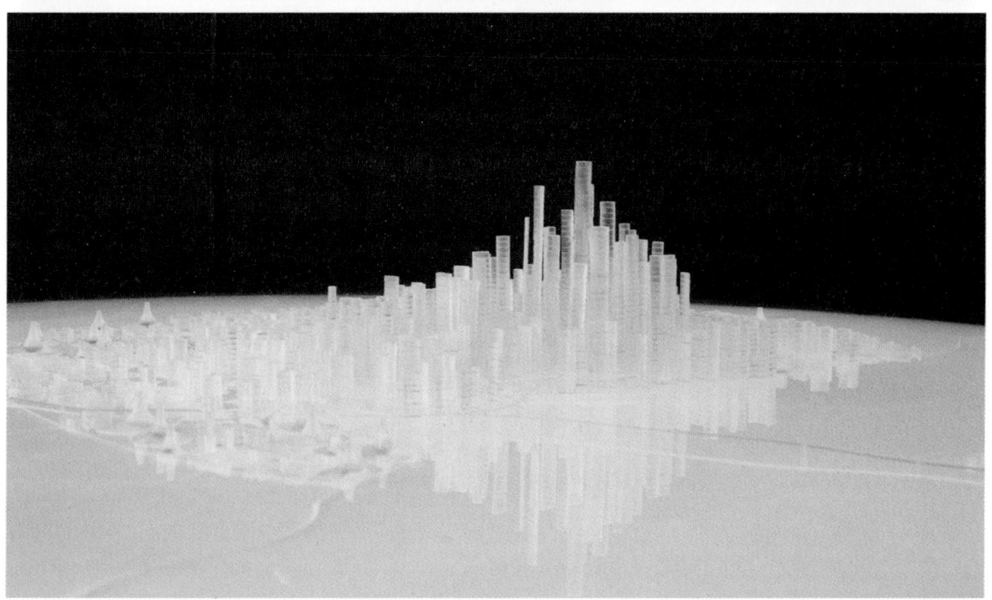

台湾塔Irma Coello + Alvaro Guinea
Taiwan Tower, Irma Coello + Alvaro Guinea
马德里网球俱乐部翻新Gil-Bartolomé建筑事务所
Tennis Club Renovation, Madrid, Gil-Bartolomé Architects
2015米兰世博会西班牙展馆
Spain Pavilion Expo Milan 2015,
萨格勒布市巴德尔街区Gonzalo del Val
Badel Block, Zagreb, Gonzalo del Val
PROFESSION
职业
墓地扩建Lorenzo Fernández-Ordoñez
Cementery Expansion, Moraleja. Lorenzo Fernández-Ordoñez
Services Building in Mayor Square, Madrid, Javier Sanjuan
Auditorium Theatre of Guadalajara, Angel Verdasco
Rental Offices In Technological Park, Zamudio, Beatriz Inglés
Arts Centre A Coruña, Victoria Acebo + Angel Alonso
Science and Technology Museum, Cuenca, Eva Hurtado
Public State Library, Santiago de Com
Prado媒体实验室Langarita-Navarro
Medialab-Prado, Langarita-Navarro Archit
Domestic Landscapes, Alcalá, Aybar-Mateos
Albacete Police Headquarters, Beatriz Matos + Alberto Castillo
Museum of the Soil, Madrid, Conchita Lapayese
Library in Colmenar Viejo, Fernando Espuelas
马德里太阳门改造CotaCero
Redevelopment of the Puerta del Sol, Madrid, CotaCero
POM Albacete, Francisco Javier González
Juan Carlos I Park, Madrid, José Luis Esteban Penelas
ACADEMIA
学术
84 Public Housing Units, Bilbao, Eduardo Belzunce
102 Dwellings in Carabanchel, Madrid,
Academie MWD Dilbeek, Carlos Arroyo

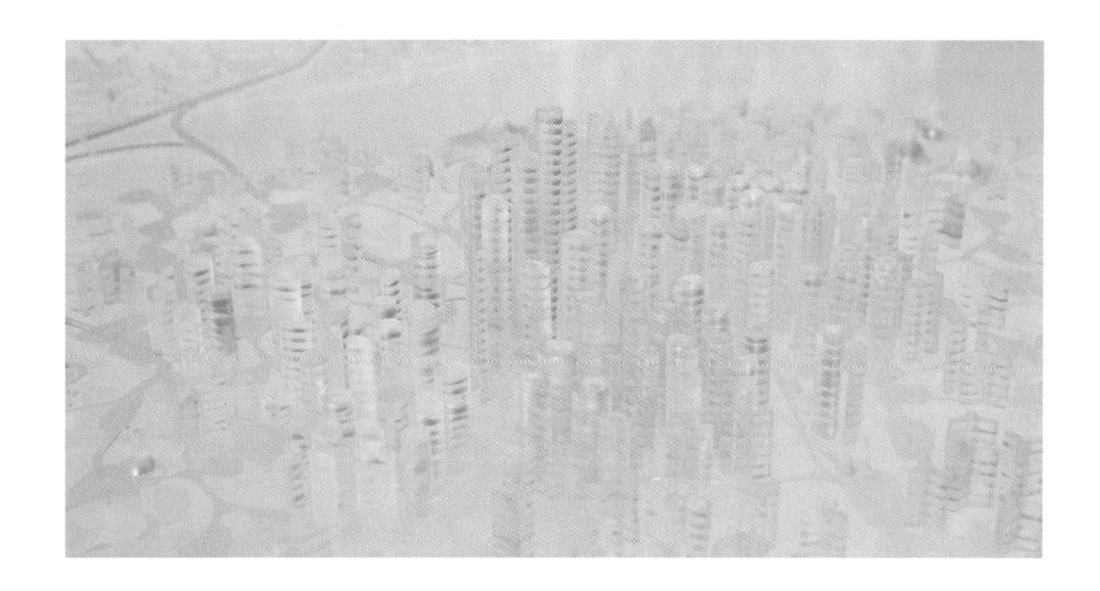

“感觉马德里”将三大主要设计理念融合于一个通用的公共空间，将其视为类似于全球网络系统一部分，提醒市民去关注它们的环境表现：未来（将设计设定为创造未来轮廓的工具）、生态与环境（未来构成的基本要素）、休闲（作为面向可持续性的生产工具）。同时，它探索六大极为重要的关键点，其在细胞生长的逻辑学基础上运用了超越计算和参数化系统的方式：改善周边、彻头彻尾的新生、社会经济的融合、战略目标、超级绿色意识、遇见趣味。

**F.E.E.L. Madrid integrates 3 strong core concepts into a generic public space as part of a global network of similar nods where citizens become aware of their environment's performance: 1. Future (setting up design as a tool to produce a tangible future); 2. Environments and Ecology (as constitutive parts of the future); 3. Leisure (as a productive tool orientated towards sustainability). It seeks to maximize 6 key drivers by using trans- computational, parametric systems, taking as its basis the logics of cellular growth: Site physical improvement - Top to bottom revitalization - Socio-economic Integration - Strategic Targeting - Super-green awareness - Meeting playfulness.**


**何塞·潘纳拉斯、佩德罗·阿尔巴、阿尔瓦罗·马丁 | Jose Penelas, Pedro Alba, Álvaro Martín**

何塞·潘纳拉斯 / 建筑师博士，首席教授，AIR Lab 设计学院院长

佩德罗·阿尔巴 / 建筑师博士，工程师博士，副教授，建筑系研究生导师

阿尔瓦罗·马丁 / 日宏（上海）建筑设计咨询有限公司，设计主管

这是一支建筑师衍生出的网状团队，包括城市规划师、社会学家、哲学家、历史学家、电影制作师、艺术家、环境学专家以及工程师。

Jose Luis Esteban Penelas/Dr Architect, Chair Professor, Director AIR Lab Design Institute

Pedro Pablo Arroyo Alba/ Dr Architect, Dr Engineer, Associate Professor, Director of the Master's Degree in Architecture (Professional)

Álvaro Guinea Martín/ Architect, Senior Architect for Nihon Sekkei (Shanghai), Inc.

This research team is composed by an extensive network of architects, urban planners, sociologists, philosophers, historians, filmmakers, artists, environmental specialists, and engineers.

深圳 SHEN ZHEN
龙岗 LONG GANG
南山 NAN SHAN
福田 FU TIAN
罗湖 LUO HU

# D-15

## 城市更新中的都市实践
## URBANUS AND THE URBAN REGENERATION IN SHENZHEN

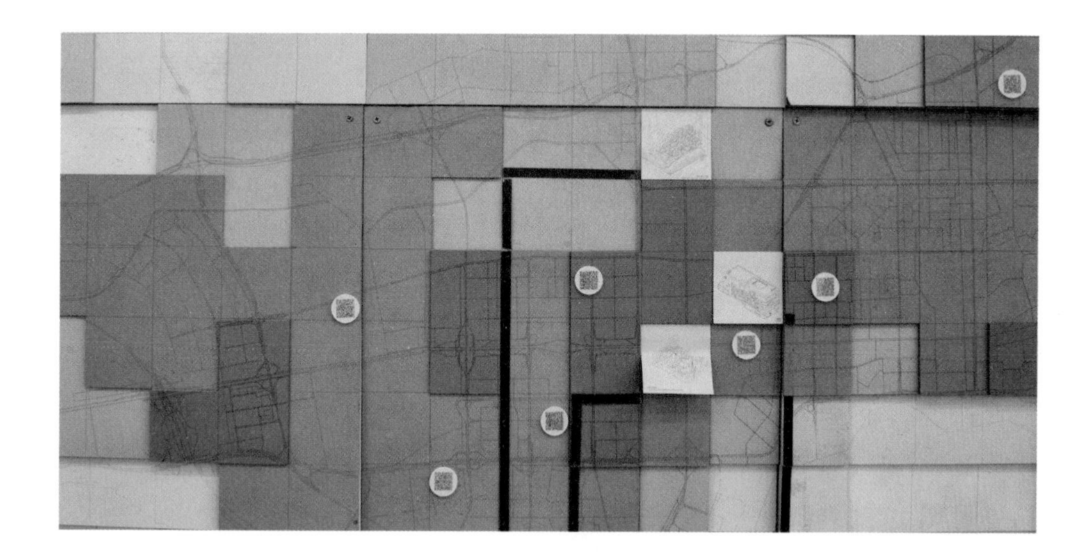

**都市实践在过去的十几年来，伴随着深圳的城市更新步伐，持续进行研究和实践，形成了一条独特的发展轨迹。从早期开始的城市公共空间营造，到对城中村改造和低收入人群的生存状态的持续关注，再到对深圳城市文化与创意产业的研究与实践，以及近来的超高密度城市更新的策略研究，都市实践始终坚持探索城市文化的创造和建筑学的意义。**

**In the past dozen years, URBANUS has developed a unique path of research and practice that continues to keep pace with the rapid urban regeneration of Shenzhen. The urban planning of public space has been a central focus for URBANUS from the beginning, and has led to in-depth research into a variety of issues. On the subject of urban village renovation, for example, the office has paid close attention to the life of low-income communities. Through projects ranging from the cultural and creative industries of Shenzhen to recent ultra-high-density urban renewal strategies, URBANUS is devoted to exploring the significant role that architecture can play in supporting urban culture.**

**都市实践建筑设计事务所 | URBANUS Architectural Design Atelier**

都市实践由刘晓都、孟岩和王辉创建于 1999 年，注重建筑的质量与批判性，力图从城市和社会角度承载更大责任，多个建成项目成为城市生活新地标，获得重要建筑奖项并在世界各地参展及出版。Urbanus was founded in 1999 under the leadership of partners Liu Xiaodu, Meng Yan and Wang Hui. It seeks to achieve a high level of design quality through a critical approach to architecture, placing great importance on maintaining responsibility from an urban and social perspective. Many projects have become new landmarks of urban life, received numerous prominent awards, and been exhibited and published world-wide.

特赞 | Tezign
科技的8个设计议题

# D-16 科技的八个设计议题 EIGHT DESIGN ISSUES IN TECHNOLOGY

**这个装置以来源于“发现特赞”社群的八段讲演视频的八个议题——数据、体验、算法、视觉、品牌、空间、虚拟、人——为原点，来试探性地提出一个数字人文语境下的设计谱系。**

**This installation consists of eight speeches from the Tezign community. They are data, experience, algorithm, visual, brand, space, virtuality, and human, respectively. It attempts to formulate a collective design discourse by bridging technology with humanities.**

**发现特赞 | Discover Tezign**

“发现特赞”是一个近十万用户的设计和科技桥接的微信公众号，隶属于特赞信息科技 (www.tezign.com)。特赞是一个让设计师自由工作的互联网平台。

范凌，设计师和互联网创业者，特赞创始人，哈佛大学设计博士。

莫万莉，建筑师和学者。耶鲁大学建筑学硕士。

Discover Tezign is a 100k-subscriber wechat public account focusing on connecting design with technology. It belongs to Tezign Tech & Design Ltd, a marketplace for high skilled creativities.

Ling Fan is a designer and entrepreneur. He is the founder and CEO of Tezign. He received a doctoral degree in design from Harvard Graduate School of Design.

Mo Wanli is an architect and a researcher. She graduates from Yale University with a master of architecture degree.

# D-17

## 上海临界地带：湿地生态盆景
## LIMINAL ZONE SHANGHAI: WETLAND BIOME BASINS

如同全球其他滨水城市，上海的河道和河口湿地也因为人类干扰正在高速碎片化和衰退的过程中。湿地作为土地和水之间的弹性缓冲地带，是极为高效的生态系统之一。湿地的集水功能、滤水功能和蓄水量让它在我们面对海平面上升和全球变暖的当下显得更为重要。湿地能控制水土流失，过滤污水，为动植物提供栖息地，并为人类提供生物多样性和美感。" 上海临界地带：湿地生态盆景 " 是让观众参与本地湿地植物滤水过程的公共互动艺术装置。克利斯曼 + 皮特鲁斯建筑事务所设计了一系列条形的湿地盆景，让市民有机会学习环境知识，欣赏植物景观，并建立起对黄浦江景观的联系。

**Along with many global waterfront cities, Shanghai's estuarine and riverine wetlands are being fragmented and degraded at an alarming rate due to increased human disturbance. As the resilient threshold between land and water, wetlands are some of the most productive ecosystems in the world. Their water collecting, filtering, and holding capacity is particularly crucial as we face rising sea levels and climate change. Wetlands control erosion, purify water, provide vital plant and animal habitat, and delight humans with their biodiversity and beauty. Liminal Zone Shanghai: Wetland Biome Basins is an interactive public art installation that viscerally engages visitors in the dynamic cycle of native plant wetland filtration. Crisman + Petrus Architects have designed a series of linear constructed wetland basins for Shanghai that create a place to learn, refresh, and conceptually connect with the nearby Huangpu River. Like the constructed wetland aboard their traveling Learning Barge project.**

**克里斯曼 + 皮特鲁斯建筑事务所 | Crisman + Petrus Architects**

克利斯曼 + 皮特鲁斯建筑事务所从事横跨建筑、景观和城市尺度的设计。他们用创新的建筑设计和城市环境修复项目探索环境、道德、美学和社会的交点。克利斯曼在弗吉尼亚大学任建筑学教授和全球可持续环境专业主管，并将她的研究和教学成果带入设计实践。 Crisman + Petrus Architects work across scales in the space between architecture, landscape and urbanism. They explore environmental, ethical, aesthetic, and social intersections through off-the-grid architecture and catalytic restoration projects for derelict and contaminated urban sites. Crisman combines practice with research and teaching at the University of Virginia, where she is a Professor of Architecture and Director of the interdisciplinary Global Environments + Sustainability major. (www.crismanpetrus.us)

# 互动

## 艺术介入公共空间

## INTERPLAY

E

## 互动：艺术介入公共空间

“城市更新”作为一种城市发展的方式，为城市以及其中的人们提供了一个关联了历史、现在和未来的时空尺度。艺术作为其意义的表征，以绘画、雕塑、装置、影像等诸多表现形式，将“城市更新”这一复杂、多维的主题进行了呈现、检省和憧憬。无论是对当前城市更新中的诸多问题的批判，还是对城市状态的创意呈现，或是对人之所向的美好愿景的描绘，“人”始终成为艺术作品关注的主体，城市的失真、冷漠、记忆和尚未可能都在与“人”的参照与比对之下，形成艺术作品最直入人心的感知路径。

## INTERPLAY: INTERVENTION OF ART INTO PUBLIC SPACE

As an urban development pattern, "urban regeneration" offers cities and city dwellers a spatial-temporal dimension that connects the past, the present and the future. Art is the representation of the connotation of "urban regeneration", as it manifests, examines and envisions this complex and multidimensional theme through the forms of painting, sculpture, installation, film and image, etc. Whether the criticism of problems in current urban regeneration schemes, or creative presentation of urban condition, or the depiction of prosperous projection, "human" is always the central concern of the artworks. The distortion, indifference, memory and possibility of the city in human scale provides the most striking perceptive path to access the artworks.

## 当代艺术理念与经验植入城市更新

张晴

30 余年快速的城市扩张为我们建立起一个巨大的表象式豪华城市景观，其间粗放不当的城市开发模式也带来了诸多无法回避的城市疾患：城市开发过度与城市管理理念及体制滞后，由此而激发的诸多社会问题，过度的资源消耗与自然环境有限承受力间的尖锐矛盾，城市空间供给过度与城市文化及情感缺失间的巨大落差，城市新生代社会文化身份的错位……这些都成为中国城市发展告别"大跃进"式的以物质空间为主导的发展模式，进入城市空间品质的新阶段伊始，我们必须正视与厘清的城市规划与建筑设计所存问题的现实。

长久以来不断加速的城市化进程成为艺术勃兴的直接缘由和内在动力，并使艺术在与国际艺术潮流碰撞的过程中实现了自身文化的确立与艺术理念的转向。今天，当我们面对野蛮生长所导致的诸多城市坎坷与误区，试图以"城市更新"作为命题，转变思路与理念、进行反思与检讨，摸索城市以及"城市人"一个关联历史、现在和未来的时空新尺度，我们更亟需艺术与文化对城市发展的洞悉与检视，拒绝长久盘踞于城市发展空洞理论覆盖下的隔靴搔痒，撕开沉醉于浅层概念移植掩护下的"掩耳盗铃"式表象，提供"城市更新"一个更为广博的视角与理解框架，一种与所在城市空间血脉相连的真实知觉。

从此意义上讲，本次展览正是在城市发展新语境下的一次可贵尝试，更是城市与艺术融汇、更新的重要转折点。其中的"互动：艺术介入公共空间"板块，也正是关注当代艺术如何以自身的理念、经验及其成果，主动地启迪"城市更新"的思想家、规划师、建筑师及相关的行业。如此，艺术"不再是单独的、孤立的现实"，而是作为"城市更新"的灵魂推动力量，植入更新城市空间与生活内涵：一方面，艺术始终站立于生活的巅峰或者彼岸，时刻以一种批判和拯救的姿态面对生活，提供城市物质建设日常之外的新鲜感知；另一方面，作为开放性、颠覆性、领导性的当代艺术，为观照和反思艺术的当代形态及城市的发展脉络与文化纬度提供了新路径，并以当代艺术的智慧链接了城市空间与人的生活方式本身。当代艺术作为"能够思考的意识"对城市空间的植入，使艺术的公共性得以进一步发酵，艺术的理念与经验由此成为引领改造城市空间与生活的重要策略，并作为城市更新实践的思想源泉、形式与语言的指针，提供了艺术与城市人关系更新的强劲动力。

在这次展览中，32 位 / 组国内外艺术家以"城市更新"为主题，以各自的艺术视角从城市问题、城市与生活、城市与生态、城市文化根植等诸多方面来反映城市社会与生活幻想的现实，揭示城市人如何主动塑造艺术植入城市空间的互动，反映了当代艺术家面对都市问题、社会问题、生活问题、经济问题及其情感问题之下，对于"城市更新"过程中诸多问题的感受力与表现力。

**城市问题的万象：**
**是不是“不能停”？**

35 年来，中国有 5 亿多人迁入城市。这种令人吃惊的城市化速度支撑起中国经济的迅猛发展，也塑造了惯性巨大的“速食”文化。“不能停”的节奏使中国人成为世界上最忙碌的群体，催生了繁荣、绚丽的城市景观，同时也亲手缔造了日益升级的城市矛盾与冲突。艺术北斗星小组的作品《世界工厂》选取服装制作业中的标志——服装加工厂现场作为主要载体，描述了催生中国巨大经济成就的粗放型中国制造业在城市更新进程中所面临的城市产业升级与人口变迁等亟待解决的重要问题。作为“世界工厂”的中国，其服装出口量占世界服装业出口总量的 33.5%、占亚洲服装业出口总量的 51.5%。而这些都是以惊人的自然资源消耗与破坏为代价的，据统计，我国每生产一件服装平均污染 0.2 立方米水，产生 0.1 立方米垃圾，消耗燃煤 1.2 千克，产生 1.13 千克碳排放。而中国出口一亿件服装才可以换回一架波音 777-300ER 飞机。当制造业面临转型，大批破败的服装制造工厂以及失业的农民工成为城市更新中最触目惊心的景观。艺术家以此引发我们对劳动密集型加工业与城市更新之间的关系的思考、对经济发展模式的反思、对未来城市的忧虑。隋建国的雕塑《中国制造》也以集装箱与文字“MADE IN CHINA”的并置方式提醒着世人：当中国已经一跃成为世界第二大经济体与世界制造业基地时，文化的贸易仍然空心化严重。据我国新闻出版总署统计的 2008 年文化产品进出口数据来看，我国图书、报纸、期刊类出口 3 487.25 万美元，进口却达到 24 061.40 万美元。音像制品及电子出版物出口为 101.32 万美元，进口为 4 556.81 万美元。版权方面，出口 2 455 种，进口 16 969 种。文化贸易中存在的巨大进出口差异清晰标示着中国文化在国际市场上的鲜明弱势，而这种城市发展与更新过程中文化的缺失是我们不能承受之重。马堡中的《不能停》也将城市空间生产的重要工具之一——水泥搅拌车作为视角。这一“城市化大生产”的典型佐证以其“不能停”的持续的运行状态，既讽刺了 30 多年来的粗暴发展方式，亦向今天的我们发出警告：“城市更新”阶段的建设是否依然“不能停”？同样，胡项城的作品《作品第十三号》用空置的透明玻璃盒子，以极简的艺术语言讽刺了城市化进程中超量城市生产导致的诸多公共空间的空置，也是中国城市发展中物质空间奢侈而文化内核干瘪的深刻现状。孙哲政的作品《无语空间》则让地锁占领停车场，作为对以汽车为主导而导致的城市交通混乱与拥堵的反抗。我们一方面在不断缔造更大更快的“一小时经济圈”的城市集群，另一方面自身城市内部的交流却要花上几倍的时间。“快城快客”每天都在遭遇着最真实的“极缓极慢”的悖论。艺术家以其敏锐的嗅觉与独特的视角，将 30 余年城市化进程中积攒的城市问题资源逐一掘出，这是我们进行城市更新的宝贵参照，亦是我们新时期前行的起点，不可回避！

**城市与生活：**
**是不是“城市灯光剧”？**

城市化是一种思维，一种组织社会的方式与经济模型的历史动力，它深刻地改变着空间与社会，也改变着中国的城市生活。作为城市更新的目的与落脚点，生活的鲜活、多样却不能仅靠物质空间的存在来确认。值得深思的是，在过往的快速城市建设过程中，太多的物质空间失去了情感的依托，呈现出错位、寂寞、失落。德国艺术家飞苹果的作品《城市灯光剧》以声控节能设施的使用不当，嘲讽了这种城市空间感受中的错位。作品讲述了装满声控灯光的城市大楼被放置在喧闹的城市街道空间之后，不但失去了原来节能的作用，反而为城市奉献了极具戏虐效果的“城市灯光剧”。对城市建筑、技术与空间之间失和的现状的深思是作品想要传达的核心内容。芬兰艺术家奥托·卡文的作品《鸟巢》则隐喻了充斥着监视器的城市，如鸟巢一般禁锢着城市人的自由，改变着城市人的生活与感知模式。岳敏君的《媒体城市》、许旭兵的《二维码》等作品则指向了城市文化产业与媒体的更新命题。媒体与传播方式的更新所造就的新语境，与新一代人思想的更替、生活方式的变迁紧密相连。媒体在平面化人们生活的同时，也异化着人的感知方式。面对媒体这一双刃剑笼罩下的城市更新新语境，我们该如何与城市重新相处，如何在错位的时空交织中放置我们的身体与情感，这正是强迫的、血淋淋的城市更新！喻红的《白领》则以少数的女性视角，描述着城市更新进程中不断蜕变的女性群体及其精神生态。城市更新中弱势女性群体的被忽视，正是城市更新需要关注的问题。王启凡的《宅城一族》、梁美萍的《城市编年史》等作品则都是对当今快捷、封闭的城市生活方式的反思。在中国都市化进程中，成长起来的“城市新生代”开始发挥着越来越重要的社会影响，与此同时，城市新生代与既有的城市社会格局之间的冲突也在不断呈现和升级，如果新一代人如此生活在虚拟世界中，城市未来的更新之路何在？艺术以其细腻的感知为城市更新延伸了触角，也不断警示我们，城市更新无法脱离生活与情感而进行单纯物质化的讨论！

**城市的未来生态：**
**是不是“一立方米风景”？**

城市从来不是规划师与建筑师的专属课题，而是囊括了生态、社会、文化等诸多维度的复合存在。当下的城市发展，极好与极坏并存。长期以来的单向发展模式使城市生态环境应有的平衡受到了严重破坏。这将成为我国在城市更新中面临的最紧迫的问题。艺术家对此进行了回应。余旭鸿的作品《一立方米风景》以废弃的矿泉水瓶为原料，为我们重塑了晶莹剔透的水晶风景，腐朽与神奇的共存让我们深刻思考美的涵义与无法回避城市的生态问题。卜冰的《细语丛林》、胡荣生的《聆听城市》则以城市中的人工材料编织着自然之幻境。作品中田园感受与人工表象间的巨大落差迫使我们去思考：我们的城市是否只能充当与自然的屏障？我们的田园情怀是否只能成为标本？

**毁灭城市文化文脉与根植的是谁：**
**是不是“推土机”？**

城市更新，不仅仅关乎建筑、空间、桥梁等物质的链接，更与城市人的生活、精神面貌是紧密相连的，是我们寻求记忆与家园的重要线索。西班牙艺术家拉拉·阿尔玛雀纪的作品《花园里的小山丘》以我们城市化进程中最常见的“推土机”为视角，思考城市更新与土地、家园、记忆之间的深刻关系，这不得不让“城市更新”的积极分子深思“推土机”是“城市更新”的英雄呢？还是毁灭城市文脉与根植的凶手？罗永进的《摩天楼》则以戏谑的手法将当代上海的标志建筑与古代青铜器相熔，形成怪诞的摩天楼形象，以此思考城市的历史与今天，是对上海与当今城市自我妖魔化的反讽。裴咏梅的《双重时间》把莫斯科的贸易、文化、生活移植到北京，形成了双重的城市生态与城市时间。胡介鸣的《家在何处》则以广州、上海这两座中国的标志性城市总图的交叠尖锐，应对着当前城市肌理大量同质化、缺乏归属感的问题。这正是对当前城市建设中缺乏创新性的例证，是不是当今中国现实版的《玩乐时间》（*Play Time*）？王维的《永恒的纪念》则以广阔空旷的内蒙家园屏蔽与抵抗着汹涌来袭、缺乏情感尺度的当代城市。这样珍贵的关于家园的记忆，还能保留多久！《百年上海城市更新中的社会生活文献与摄影项目》则以百余张的黑白照片，再现了上海自开埠以来的前世今生。何为我们的城市？我们的城市要"望得见山，看得见水，记得住乡愁"。正如许江作品《盘根》中以遒劲的秋葵之根所警示的：文化与记忆的根植永远是城市更新的灵魂，是我们寻回家园的重要精神维度所在！

城市问题的万象：是不是"不能停"？
城市与生活：是不是"城市灯光剧"？
城市的未来生态：是不是"一立方米风景"？
毁灭城市文化文脉与根植的是谁：是不是"推土机"？

这些都是艺术家针对"城市更新"的反思、批判与扪心质疑。这些问题既是"城市人"的思考、建筑，城市规划等专业人士的思考，也是艺术家的思考，我们任何人都无法置身于外。当艺术家以绘画、雕塑、装置、影像等诸多表现形式，将"城市更新"这一复杂、多维的主题进行呈现、省思、协商时，艺术的当代性植入到更普遍的城市空间之中，为城市空间的更新与再造提供了新的思想对话框架与视觉美学。而艺术与建筑、规划等形式被共同运用于城市空间的再造中，其方式也营造出了一种超验的对话性创作，使艺术以"当代性"作为颠覆的切入口，站在他者的位置，重新审视我们熟悉的空间，体验一种身体的顿悟。城市的参与者与日常使用者一起被推出共识语境、以再现方式和每个人自我意识的熟悉连接，使具去质询眼前的城市规划与建筑设计尴尬的景观、程式化图像与空间的叹惋。

当代艺术理念与经验植入城市更新，充当着城市创意的想象力及其推动力，亦提供了反观城市更新的崭新维度：无论是对当前"城市更新"中的诸多问题的批判，还是对都市状态的创意，或是对城市人所向往的美好愿景讨论，"人"始终成为艺术作品关注的主体。城市的冷漠、失真、记忆和未知可能都在与"人"的参照与比对之下，形成"艺术更新"最直入人心的感知路径！

## Implantation of Contemporary Art Concept and Experience into Urban Regenaeration

**Zhang Qing**

With the rapid urban expansion for more than 30 years, we have established a huge representation of luxurious urban landscape, but the improper extensive urban development has also incurred numerous unavoidable urban diseases: lots of social problems caused by the excessive development and the backward concepts and systems of urban management, sharp contrast between the unreasonable resource consumption and the limited environmental endurance, great difference between undue supply of urban spaces and urban culture and athymia, as well as the misplacement of the new-type urban social and cultural identities. All these diseases initiate the beginning for Chinese urban development to get rid of the "great-leap-forward" development mode based on material space and enter into a new phase of urban space quality, so we have to envisage and clear up the reality of problems existing in urban planning and architectural design.

The constantly accelerating urbanization process has long become the direct reason and internal power of art, and enables art to realize the establishment of its own culture and the diversion of art concept during the collision with the international art trends. Nowadays, in face of various urban difficulties and misunderstandings resulted from barbarous growth, when we attempt to convert our ideas and concepts, conduct introspection and self-criticism based on the theme of "urban regeneration" and explore a new space-time dimension of city and "city people" relating to the history, present and future, it is more necessary for us to understand and view urban development via art and culture, reject the long-term vain attempts of void urban development theories, tear the literal appearance covered by the superficial concept implantation and provide a broader viewpoint and understanding framework for "urban regeneration", as well as a real perception closely connected with the urban space.

Viewed from such a meaning, this exhibition is a valuable attempt under the new context of urban development, and even an important turning point for the integration and upgrading of city and art. In the exhibition, the "Interplay: Intervention of Art into Public Space" section exactly focuses on how the contemporary art enlightens actively the ideologists, planners, architects and related industries of "urban regeneration" via its concepts, experience and outcomes. Thus, art is "not a lonely and isolated reality", but, as the central driving force of "urban regeneration", is implanted into the renewal of urban space and life connotation. On one hand, art is always on the peak of life, facing with life in a critical and saving manner and provides fresh perceptions beyond the material urban construction; on the other hand, contemporary art with openness, subversiveness and leading offers new paths to observe and reflect the contemporary form of art, as well as the urban development vein and cultural dimension, and links with urban spaces and human's lifestyle with the wisdom of contemporary art. Contemporary art, as the implantation of the "consciousness capable of thinking" into urban spaces, makes the further fermentation of the artistic publicity so that the art concept and experience become an important strategy to lead the transformation of urban space and life and, as the pointer of thinking source, form and language to practice urban regeneration, provides strong power for renewal of relationship between art and city people.

In this exhibition, 32 (groups) of domestic and international artists, from their own artistic perspectives in various aspects such as urban problems, city and life, city and ecology, urban culture rooting, etc., with the theme of "urban regeneration", reflect the reality of urban society and life illusion, reveal how the city people actively shape the interaction of art implanted into urban spaces and indicates the contemporary artists' sensibility to an expression of numerous problems in the process of "urban regeneration" in face of urban, social, life, economic and emotional problems.

## Can urban problems be *No Stop*?

For thirty-five years, more than 500 million people have moved to urban areas in China, which represents an astounding urbanization speed, supports the rapid economic development of China and shapes the great-inertia "fast food" culture. Such rhythm of "incapability to stop" enables Chinese to become the busiest population in the world, forces the flourish and dazzling urban landscapes and also brings about the increasingly upgrading urban contradictions and conflicts. In *World Factory*, Beidou Art Group selected the sign of the clothing manufacturing industry-the site for clothing manufacturing, as the main carrier, describing such key problems necessary to be solved in the urban regeneration process of the extensive Chinese manufacturing industry which gives birth to enormous Chinese economic achievements, such as urban industrial upgrading and demographic change. As the "world factory", China has the clothing export volume accounting for 33.5% of the total export volume of the world's clothing industry and 51.5% of the Asian clothing industry, which is at the cost of vast consumption and destruction of natural resources. Statistics show each piece of clothing manufactured in China consumes $0.2m^3$ of water, produces $0.1m^3$ wastes, consumes 1.2kg of fuel coals and generates 1.13kg carbon emission. What's more, 100 million pieces of clothing exported by China can exchange a Boeing 777-300ER aircraft. When the manufacturing industry is going to be transformed, a large batch of run-down clothing manufacturing plants, as well as the unemployed peasant workers will become the most startling landscapes in the urban regeneration. Thus, the artists trigger us to think the relationship between the labor-intensive manufacturing industry and urban regeneration, introspect the economic development model and worry about the future city. Sui Jianguo, in his statue *Made in China* also, via both the containers and the words "MADE IN CHINA", reminds people that when China has become the world's second largest economy and the world's manufacturing base, the cultural trade is still serious in hollowing. The General Administration of Press and Publication of the People's Republic of China (GAPP) made statistics of the cultural product import & export data in 2008, according to which, China achieved the export and import volumes of respectively USD 34.8725 million and USD 240.614 million as for books, newspapers and periodicals, and of USD 1.0132 million and USD 45.5681 million as for audio-visual products and e-journals. Regarding to copyrights, it exported 2,455 kinds while imported 16,969 kinds. The great difference between import and export in cultural trade clearly indicates the evident weakness of Chinese culture in the international markets, however, such kind of culture default in the process of urban development and regeneration is our unbearable burden. Ma Baozhong, in his *No Stop*, also selects the cement mixer, one of the important tools produced by urban spaces, as the perspective. The typical evidence of such "urbanized mass production" mocks the rude development pattern for more than 30 years through the unceasing operation status, and proposes a question to us today that whether the construction in the "urban regeneration" stage cannot still be stopped. Similarly, adopting the empty transparent glass box and extremely simple art language, Hu Xiangcheng, in his *No. 13*, satirizes the numerous vacancies of public spaces caused by excessive urban production in the process of urbanization, which is also the in-depth status of extravagant material space and wizened cultural core during Chinese urban development. Sun Zhezheng, in his *Silent Space*, describes the ground locks spread over the parking lot and resists the urban traffic disturbance and congestion mainly incurred by automobiles. We unceasingly create the megalopolis of large and faster "one-hour economic circle" but still spend severalfold time on urban traffic internally. "Trans local motion" suffers from the truest paradox of "extreme slowness" every day. The artists, through their sharp senses and unique perspectives, represent various urban problem resources accumulated in the urbanization process for more than 30 years detail by detail, forming the precious reference and consciousness for us to carry out urban regeneration, as well as our starting point to move forward in a new era. All these problems can't be ignored.

## Are city and life a *Licht Spiele*?

As a kind of thinking, a way to organize society and a historical driving force of economic model, urbanization deeply changes spaces and society, as well as Chinese urban life. The freshness and diversity of life are the destination and foothold of urban regeneration but can't be defined only by existence of material spaces. What is worth pondering is that in the process of previous fast urban construction, too many material spaces lost the basis of emotion, but presents misplacement, loneliness and disappointment. German artist Alexander Brandt, in his *Licht Spiele* via improper utilization of sound-controlled energy-saving facilities, sneers the misplacement in such urban space perception by depicting that the urban buildings fully equipped with sound-controlled lights, after being placed into the noisy urban street spaces, fail to play their original roles of energy conservation but present a very joking "urban light play" for the city, mainly expressing the deep consideration of the current situation of imbalance among urban buildings, technologies and spaces. Finish artist Otto Karvonen, in his *Shanghai Birdhouses*, displays a metaphor that the city filled with monitors, like a bird's nest, imprisons the freedom of city people and changes the city people's life and perception modes. In Yue Minjun's *The Media City* and Xu Xubing's *D Bar Code*, etc., the subject of renewal of the urban cultural industry and media is focused on. The new context created by the renewal of media and transmission methods is closely linked with the ideological replacements of a new generation and the changes of lifestyles. The media brings planarization for people's life and simultaneously, alters their perception methods. In face of the new urban regeneration surrounded by media as a double-edged sword, it is the forceful and bloody urban regeneration that how we deal with the city and place our bodies and emotions in the misplaced space-time intertexture. Yu Hong's *White Collar* depicts the constantly variable females and their spiritual ecology in the process of urban regeneration from the rare female perspective. The uncaring and ignorance of female group in the urban regeneration are exactly the issue which needs attention. Similarly, Wang Qifan's *Housing City*, Leung Mee-ping's *City Chronicles*, etc. are the introspection of rapid and closed urban lifestyle. In the process of Chinese urbanization, the growing "new cities" begin to play an increasingly important role in the society, and at the same time, continuously presents and upgrades conflicts with the existing urban social pattern. If the new generations live in such a virtual world, then where is the road to the urban regeneration in the future? The art with its fine perception extends the tentacles for the city and warns us again and again that the urban regeneration is used for the purely materialized discussion which is unable to be separated from life and emotion.

## Future ecology of the city: is it a *Cube Scenery?*

City is never a special field only for city planners and architects; It's a multi-dimensional issue regarding fileds like ecology, society and culture. Two opposite extremes coexist today. One-way development mode has seriously damaged the due balance of urban ecological environment for a long term, which shall be the most urgent problem facing urban regeneration in China. Artists have made their responses hereto. In his work *Cube Scenery*, Yu Xuhong uses waste mineral water bottles to rebuild a crystal clear scenery, forcing us to think about the esthetical connotations and unavoidable urban ecological issues profoundly by way of the co-existence of both the decayed and the beautiful. In *Whispering Woods* by Bu Bing and *Listening* by Hu Rongsheng, artificial materials are used to weave a dreamland of nature. Among those works, the enormous gap between pastoral experiences and artificial images compels us to consider that: whether can our cities only act as barriers towards nature? Whether can our pastoral emotions only become specimens?

## Who destroyed the urban culture contexts and roots: is it *the Bulldozer*?

Urban regeneration not only refers to material links such as architecture, spaces and bridges, and what's more, it is closely connected to the life and spiritual outlooks of urban citizens, as well as an essential clue for us to search for memories and homelands. *The Buried House* by Spanish artist Lara Almarcegui is based on the point of view of the most common "bulldozer" in the process of urbanization, and reflects on the profound relationship between urban renewals and lands, homelands and memories, which makes the activists of "urban regeneration" have to think about whether the "bulldozer" is the hero of "urban regeneration", or the murderer to destroy urban contexts and roots. In *Skyscraper* by Luo Yongjin, landmarks in contemporary Shanghai are fused together with ancient bronze wares to become an uncanny skyscraper image in a jocosity technique, hereto reflect on histories and today of cities, which is an irony of self-demonization to both Shanghai and current other cities. In *Dual Time* by Pei Yongmei, trades, culture and life in Moscow are transplanted into Beijing, forming a double urban ecologies and time. *Where Is Home* by Hu Jieming depicts overlapping of general drawings of Guangzhou and Shanghai that are two emerging symbolic cities in China, and pointedly faces problems of mass homogenization and lack of a sense of belonging for urban textures at present. This is exactly an example of a lack of creativities for current cities construction. And is it a realistic edition of *Play Time* in current China? In *Eternal Memory* by Wang Wei, the vast and broad Inner Mongolia homelands are shielding and resisting the raging and incoming wave of contemporary cities that are lack of emotional scales. How long such valuable memories of homelands can preserve? In *Historical Changes in Shanghai*, more than one hundred black-and-white photographs reproduce the past and the present of Shanghai since its port opening. As is warned in *The Roots* by Xu Jiang, both culture and memories shall be rooted in the soul of urban regenerations all the time, as well as be the place where the important spiritualdimension is located for us to recover our homelands!

Can urban problems be *No Stop*?
Are city and life a *Licht Spiele*?
Future ecology of the city: is it a *Cube Scenery*?
Who destroyed the urban culture contexts and roots: is it *the Bulldozer*?

These are the artists' reflections, criticisms and queries on "urban regeneration". They are thought both by "urban citizens" and by professionals such as in architecture and urban planning areas, as well as by artists. None of us can stand aside. When artists present, reflect, and negotiate such complicated and multi-dimensional topic of "urban regeneration" in various patterns of manifestation such as paintings, sculptures, equipment and images, contemporaneity of art is then transplanted into more universal urban spaces, to provide new thoughts dialogues frames and visual aesthetics for the renewals and recreations of urban spaces. However, common applications of art, architecture, planning and forms into the recreation pattern of urban space also builds a kind of transcendent dialogism creation, and makes art use "contemporaneity" as the subversive entry point, which means to review our familiar space completely again and experience a sort of physical insight, standing on the side of the other.Both city participants and daily users are pushed out of the consensus contexts and link to the familiar self-consciousness of each person by reproduction, in order to query and sign in sympathy the awkward landscapes, stylization images and spaces of urban planning and architectural designs in the front.

Contemporary art concepts and experience are implanted in urban regeneration, and act as the imagination and driving forces of city creativities, as well as provide a brand new dimensionality to reflect on urban regeneration: whether it is the criticism to multiple problems in current "urban regeneration", or the discussion about the metropolis creativities or fine vision that urban citizens are yearning for, mankind shall be the subject that art works pay attention to from the beginning to the end. And any indifference, distortion, memory and unknown possibility of cities will become the most direct perception access to popular feelings for "art regeneration" in the reference and comparison of human!

# E-01

## 灯光剧
## LICHT SPIELE

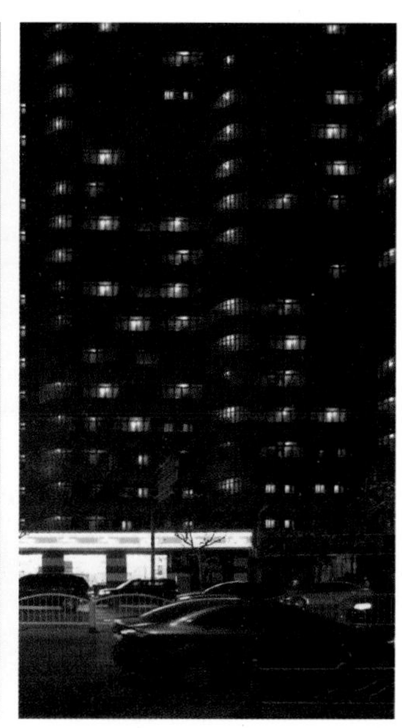

**类型 | Type: 影像 | Video**

**创作年份 | Year: 2015**

**实景纪录一段影像。十字路口不同强度的声音，影响楼房声控灯的亮灭，从而上演一出好玩的灯光剧。每一个窗口皆是舞台布景的一部分，你不知它何时亮，或者你正在期待它灭。剧情，你既可以预料，也很难完全预料。每一个路过者既是观众，也可能无意间成了幕后工作者的一员。**

**There is a building in Shanghai that is performing a very special light-show for free every night. I did nothing but film this performance.**

**飞苹果 | Alexander Brandt**

飞苹果 (Alexander Fabian Brandt)，德国新媒体艺术家，德国上海飞来飞去新媒体展示设计公司总裁兼创意设计总监。

Prof.Alexander Brandt, New media artist, Creative Design Director, Fei-Lai-Fei-Qu Multimedia Co.Ltd., Germany.

# E-02 新家园 NEW HOME

**类型 | Type: 影像 | Video**

**创作年份 | Year: 2015**

**城市是为人而建，在建城过程中人类驱赶了原本生存在土地上的万物生灵。既然没有人愿意来到康巴什新城，这片土地上的原住民——狼，回来了，它孤独地走在城市的边缘，看着这座无比熟悉的“新家园”。**

**The city is constructed for the human, during the process of construction, human beings expelled all the original creatures living in the land. Since no one is willing to come to Kangbashi, the original inhabitant of the land, the wolves, came back. They walk at the edge of the city lonely, looking at the familiar "New Home".**

**唐应山 | Tang Yingshan**

1964 年生于南京，现居北京。1990 年毕业于中央美术学院壁画系；现为中央广播电视大学教授，也是九三学社社员。Born in Nanjing in 1964, Tang now lives in Beijing. He graduated from the mural Department of China Central Academy of Fine Arts in 1990. He is Professor of The Open University of China, and a member of the Jiu San Society.

# E-03 细语丛林 WHISPERING WOODS

类型 | Type: 综合材料 | Multimedia

创作年份 | Year: 2010

尺寸 | Size: 200cm × 400cm

媒材 / 技术 | Mediums/Technology: PC 管 | PC pipe

在黄浦江畔，夜色阑珊之时，由 5000 根左右 2.4 米高的杆件构成的模拟丛林，当人接近时触发红外感知，并将信号逐个传输至下一个灯杆，像烽火台一样沿着江岸把信号缓慢传输下去，并形成干涉纹样的变化。观众的心情在灯光闪烁下，与色彩斑斓的城市生活互动。

The work consists of 5000 pieces of poles, each pole is 2.4m high such like simulated jungle. People could interact with the flashing lights to experience the colourful city life.

**卜冰 | Bu Bing**

集合设计主持设计师、艺术家，现居上海。 The chief designer of ONE DESIGN INC, artist. He lives in Shanghai.

# E-04 聆听城市 LISTENING

类型 | Type: 声音装置 | Sound Installation

创作年份 | Year:2015

尺寸 | Size : 可变 | Variable

**作品将视角集中于城市的声音，脱离了物质空间、形体的参照，并以此作为观察城市更新的独特通道。**

**The work focuses on the sound of city. Out of the physical space and the shape as a reference, it is a unique way to observe the city.**

**胡荣生 | Hu Rongsheng**

艺术家，生活与工作在苏州。 The artist now lives and works in Suzhou, Jiangsu Province.

# E-05 家在何处 WHERE IS HOME

类型 | Type: 绘画 | Painting

当城市越来越雷同，新的空间快速地将历史记忆吞噬，哪里似乎都是我们的家，又不是我们的家。艺术家以重置的方式为我们揭示了这一现实。

The cities become more and more similar. History was forgotten by people. Everywhere seems like home, except they are not. The artist shows this reality in the way of reset.

**胡介鸣 | Hu Jieming**

艺术家，生活、工作于上海。 The artist lives and works in Shanghai.

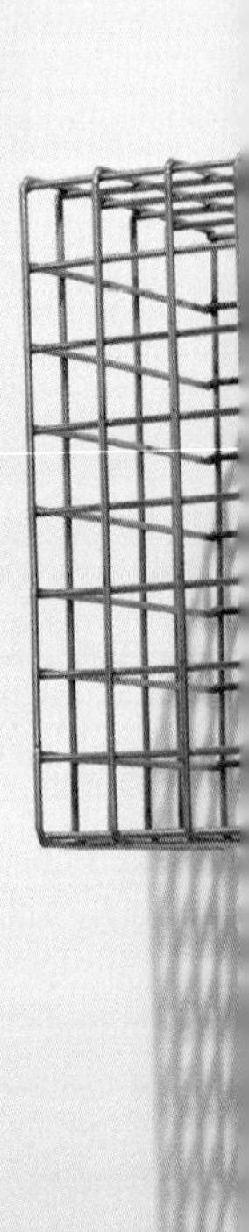

# E-06

## 上海的鸟巢
## SHANGHAI BIRDHOUSES

类型 | Type: 装置 | Installation

尺寸 | Size: 可变 | Variable

在上海将这些鸟巢设置在那些已经被新建筑所占领的老建筑原来的地方。这些鸟巢可以被放在展厅内外，以增加这些地方的历史感。鸟类是向往自由的物种，但同样不希望随意改变它们的生活环境，同时，鸟类仍会在世界范围内迁徙，以寻找最好的栖息地。但愿上海黄浦江畔飞鸟云集。

**To create birdhouses that resemble the old buildings that have been removed to make space for the new buildings. The birdhouses could be installed on the walls of the new buildings, to add a layer of history in the place. Birds are a common metaphor for freedom but they are also very adaptable to changes in their surroundings. Birds also migrate in great numbers around the world, to find the optimal habitat.**

**奥托·卡文 | Otto Karvonen**

生于 1975 年，生活和工作于荷兰的赫尔辛基。擅长于公共行为艺术与雕塑装置。他的作品经常以诙谐或讽刺的方式，诠释我们对日常现实的感知。Born in 1975, Karvonen now lives and works in Helsinki. He is good at public behaviour art and sculpture device. His work is full of humorous or sarcastic manner to express the perception of reality.

# E-07 一立方米风景 CUBE SCENERY

类型 | Type: 装置 | Installation

创作年份 | Year: 2015

尺寸 | Size: 100cm × 100cm × 100cm

媒材 / 技术 | Mediums/Technology: 水晶 | Crystal

作品直面当代生活的现场，在朴素的个体经验中获得“自然”的感受，可见心灵的山水。切取一立方米城市生活遗弃物、日常生活的剩余物，转化为 3D 水晶激光内雕装置、影像，通过镜子的反射，形成一立方米的实景和虚影，呈现出可变的山水，山水为空、静意境。

**The work shows a variable landscape. The landscape is empty and the mood is quiet.**

**余旭鸿 | Yu Xuhong**

1975 年生于浙江开化。2000 年毕业于中国美院油画系，获学士学位；之后先后于 2003 年和 2008 年获该校的硕士学位和博士学位。现工作、生活于杭州。 Born in Kaihua, Zhejiang in 1975, Yu studied at China Academy of Art, Oil Painting Department, and got his Bachelor degree in 2000, Master degree in 2003, and PhD. in 2008. He Now works and lives in Hangzhou.

# E-08

## 光影·山水
## LANDSCAPE OF LIGHT AND SHADOW

**类型 | Type: 布面油画 | Canvas oil painting**

**创作年份 | Year: 2015**

**尺寸 | Size: 450cm × 200cm, 300cm × 80cm, 150cm × 150cm, 300cm × 80cm**

**城市的飞速发展使人们习惯了生硬的物质与材料堆砌的空间，风景成为一种奢侈。作品以梦幻般的风景表达的正是城市更新中以人为尺度的追求和向往。**

**The scenery become a luxury thing during the urban development. The artist displays fantastic scenery to express a kind of pursuit and longing of people in the urban regeneration.**

**余旭鸿 | Yu Xuhong**

1975 年生于浙江开化。2000 年毕业于中国美院油画系，获学士学位；之后先后于 2003 年和 2008 年获该校的硕士学位和博士学位。现工作、生活于杭州。 Born in Kaihua, Zhejiang in 1975, Yu studied at China Academy of Art, Oil Painting Department, and got his Bachelor degree in 2000, Master degree in 2003, and PhD. in 2008. He Now works and lives in Hangzhou.

963
F2655
CAT

# E-09

## 花园里的一个小山丘
## THE BURIED HOUSE

**类型 | Type: 影像 | Video**

**作品记录了在城市更新中一座房子的拆除与掩埋，以此探讨城市与街道、城市与建筑、城市与居民之间的关系。**

**The work recorded the process of removal and burial of a house in order to explore the relationship between the city and the street, architecture and people.**

**拉拉·阿尔玛雀纪 | Lara Almarcegui**

1972 年出生于荷兰鹿特丹，于 Cuenca 大学和阿姆斯特丹的 De Ateliers 学习美术，现居住于荷兰港市鹿特丹。 Born in Rotterdam, Holland in 1972, Almarcegui studied in the University of Cuenca and De Ateliers. She now lives in Rotterdam.

# E-10 世界工厂 WORLD FACTORY

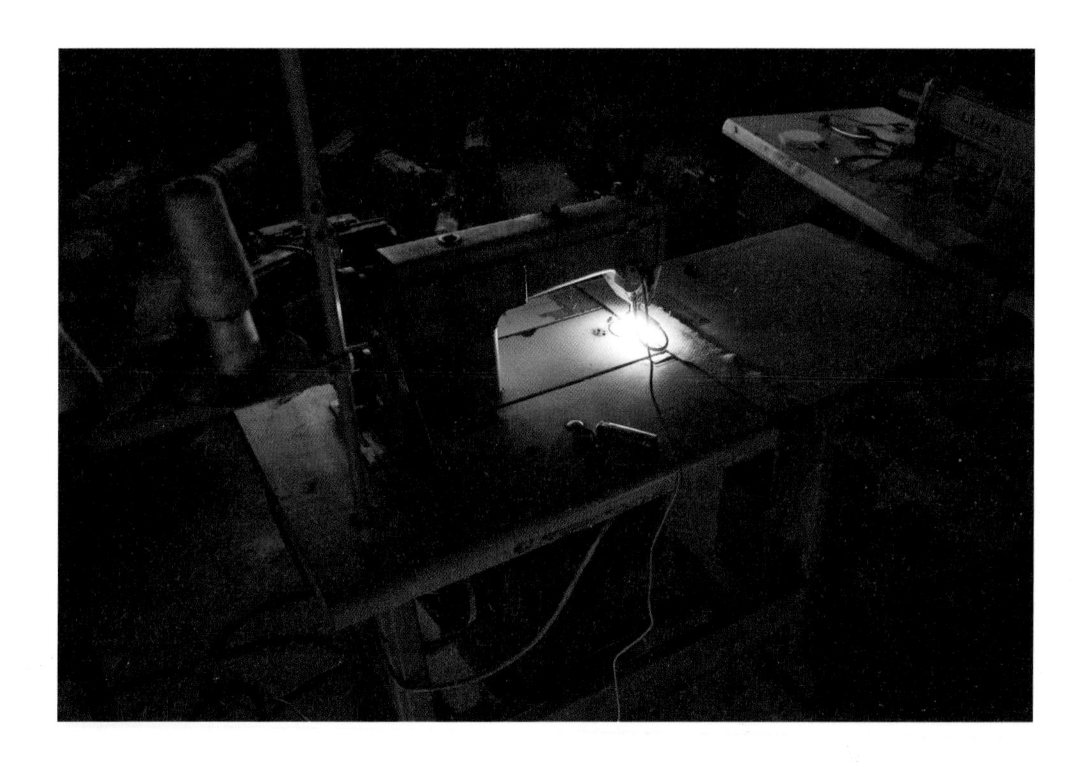

**类型 | Type: 影像 / 装置 | Video/Installation**

**作品选取服装制作业中的标志性工具——缝纫机，作为主要载体，呈现出劳动密集型产业在产生价值过程中所引发的遐想，破旧的机器上残留的世界名牌标志，隐喻出曾经为欧美服装订单而夜以继日工作的工人疲惫的身影，杂乱的碎布剩线直射出对工人身心的摧残，从而引发我们对劳动密集型加工业与城市更新之间的关系的思考、对经济发展模式的反思、对未来城市的忧虑。**

**The artists selected iconic tools of the clothes making industry, sewing machines, so as to lead us to think about the relationship between labour intensive processing industry and urban renewal, the reflection on the economic development model, and the worry about the future of the city.**

**北斗星小组 | Big Dipper Group**

刘越（上海）、吴黎中（北京）、唐应山（北京）、日青（北京）、严逢林（上海）、许旭兵（上海） Members of The Big Dipper Group include: Liu Yue (based in Shanghai), Wu Lizhong (based in Beijing), Tang Yingshan (Beijing), Ri Qing (Beijing), Yan Fenglin (Shanghai), Xu Xubing (Shanghai).

# E-11 永恒的记忆 ETERNAL MEMORY

**类型 | Type: 绘画 | Painting**

**创作年份 | Year: 2015**

**尺寸 | Size: 100cm × 100cm**

**媒材 / 技术 | Mediums/Technology: 丙烯综合材料 | Polypropylene composite materials**

**童年的记忆是我创作的源泉。荒滩，野草，“干打垒”，以及至今仍矗立在那片盐碱滩上的、已故家父设计建造的水塔和俗称“马脊梁”的苏式营房……它们填满了我儿时的记忆，形成我现今画面的元素与语言。**

**The memory of my childhood is the source of my creation. Wasteland, weeds, “dry off”. The tower and Soviet barracks which were designed by my late father are still standing there. My childhood memories were full with them which are the elements and language of my works.**

**王维 | Wang Wei**

1966 年生于北京军区内蒙古生产建设兵团二师十四团，1992 年毕业于中央美术学院，获学士学位。现居北京，职业艺术家。作品被中国美术馆、民生现代美术馆、当代华人美术馆及国内外私人收藏。 Wang was born in Beijing Inner Mongolia production and Construction Corps two division fourteen regiment in 1966. In 1992, he graduated from China Central Academy of Fine Arts with his Bachelor degree. He now lives in Beijing as a professional artist. His works have been collected by National Art Museum Of China, Ming Sheng Art Museum and a number of private collectors.

# E-12 百年上海城市更新中的社会变迁摄影文献项目 HISTORICAL CHANGES IN SHANGHAI

类型 | Type: 照片 | Photograph

创作年份 | Year: 2015

尺寸 | Size: 可变 | Variable

作品以一组珍贵的上海老照片，梳理了百年上海的社会及城市空间变迁。对于当今而言，这是最好的城市更新的例证。

**A series of rare photos of old Shanghai, traces the development and shifts of social and urban spaces in Shanghai. Today, they are the best witness of urban regeneration.**

**城市影像研究与实践项目小组 | Urban Image Research and Practice Project Team**

雨园，摄影项目组代表，生活与工作在北京。 Yu Yuan, Project representative, now lives and works in Beijing.

# E-13

## 摩天楼
## SKYSCRAPER

**类型 | Type: 雕塑 | Sculpture**

**创作年份 | Year: 2005 — 2008**

**尺寸 | Size: 95cm × 115cm**

**媒材 / 技术 | Mediums/Technology: 青铜 | Bronze**

**作品将视线投向了建筑，冷峻写实的背后，蕴涵着艺术家对老城、传统、过去以及时代更替的复杂心情。**

**The work focuses on the building, behind the cruel reality, the artist wants to express the complex mood for the old city, the tradition, the past and the times.**

**罗永进 | Luo Yongjin**

1960 年出生于北京；1978-1982，解放军外语学院，文学学士；1982-1984，解放军外语学院电教中心；1985-1986，浙江美术学院油画系进修；1988-1992，广州美术学院，艺术硕士；1993-1997，解放军外 语学院电教中；1998-1999，中央电视台美术星空，摄像师、自由摄影师；2000 年，中国美术学院上海艺术设计学院，副教授。1960, born in Beijing.1978-1982, PLA University of Foreign Language, Bachelor of Arts.1982-1984, PLA University of Foreign Language, Audio and Video Centre. 1985-1986, studied in China Academy of Art, Department of Oil Painting. 1988-1992, Guangzhou Academy of Fine Arts, Master of Arts. 1993-1997, worked at the Audio and Video Centre of the PLA University of Foreign Languages. 1998-1999, "Art World" CCTV, photographer. 2000, The China Academy of Art, Shanghai Art Design Institute, Associate Professor.

# E-14 黄浦江畔 HUANGPU RIVER

**类型 | Type: 雕塑 | Sculpture**

**创作年份 | Year: 2002—2003**

**尺寸 | Size: 1200cm × 50cm × 8cm**

**媒材 / 技术 | Mediums/Technology: 青白瓷 | White porcelain**

**艺术家以黄浦江畔为视角，用艺术的手法表现了上海这座城市最具代表性的城市公共空间的特质。**

**The artist selected the Huangpu River as the perspective to display the characteristics of the urban public space which is the most representative view of the Shanghai city.**

**刘建华 | Liu Jianhua**

1962 年出生于中国江西省吉安市，现生活、工作于中国上海。 Born in Ji'an, Jiangxi province in 1962, the artist now lives and works in Shanghai.

# E-15 盘根 THE ROOTS

**类型 | Type: 雕塑 | Sculpture**
**创作年份 | Year: 2009**
**尺寸 | Size: 可变 | Variable**
**媒材 / 技术 | Mediums/Technology: 铜 | Copper**

**那葵与大地同体同色，风烧火燎一般，熠熠然闪着铜光。那葵的极盛和衰老，只在秋夏之间。眼见到的却是废墟般的庄重。生命如此倏忽，却又要在原野上守候着自己，守候一场辉煌的老去。**

**The topic of the *The Roots* is about the questioning and calling for the cultural life.**

**许江 | Xu Jiang**

中国美术学院院长、教授，全国人大教科文卫专门委员会委员，全国文联委员，中国美术家协会副主席，中国油画艺术委员会委员，中国油画学会主席，浙江省文联主席，浙江省美术家协会主席。 President and Professor of the China Academy of Art; A member of a special committee of science education of National People's Congress; The National Federation of literary and art circles; Vice chairman of Chinese Artist Association; Member of China oil painting art committee; Chairman of Chinese oil painting society; President of Zhejiang Province Federation; Chairman of Zhejiang artists association.

# E-16
# E-17

**双重时间**
**DUAL TIME**

类型 | Type: 绘画 | Painting

KAPГO
KASTORIA

类型 | Type: 录像（四屏录像 | Four-screen video）

创作年份 | Year: 2012—2015

尺寸 | Size: 630cm × 250cm, 400cm × 250cm, 400cm × 250cm, 500cm × 220cm

媒材 / 技术 | Mediums/Technology: 布面油画 | Canvas oil painting

作品《双重时间》是全景式地展示当下俄罗斯文化注射在北京大众文化场所的缩影，作品由绘画、装置、录像、摄影、声音、灯光及综合艺术构成于一个通透的空间，作品将现成品的装置现场及其富有极强表现力的绘画形成交相辉映的陌生感，以 Art Deco 样式的阶梯引领人们的视线进入一个具有仪式感的光韵与镜像世界，把匿名性的悸动推向无可预测的再造。作品《双重时间》就是淋漓尽致地展示当下的景观世界，重新启动不可复制的和可复制的幻异未来。

The work *Dual Time* can be regarded as a panoramic view of the recreational facilities in Beijing, which are under the direct influence of Russian culture. The combination of art forms constitutes a transparent space, including drawing, installation, video, photography, sound, light, and synthetic art. The ready-made installation and expressive drawings enhance each other's beauty and a sense of 'unfamiliarity'. The Art Deco style of ladder brings the audience into a ritualized mirror world of light, pushing the anonymous sensation into an unpredictable re-making. *Dual Time* most beautifully describes the current world, initiating a vision of the future - the (un) duplicable fantasy.

**裴咏梅 | Pei Yongmei**

1975 年于天津出生，现工作与生活在北京，并任教于中央美术学院油画系第四工作室，副教授。 Born in 1975 in Tianjin, China, Pei now Lives and works in Beijing. She is teaching at Oil Painting Department,Central Academy of Fine Arts as an Associate Professor.

# E-18 二维码 2D BARCODE

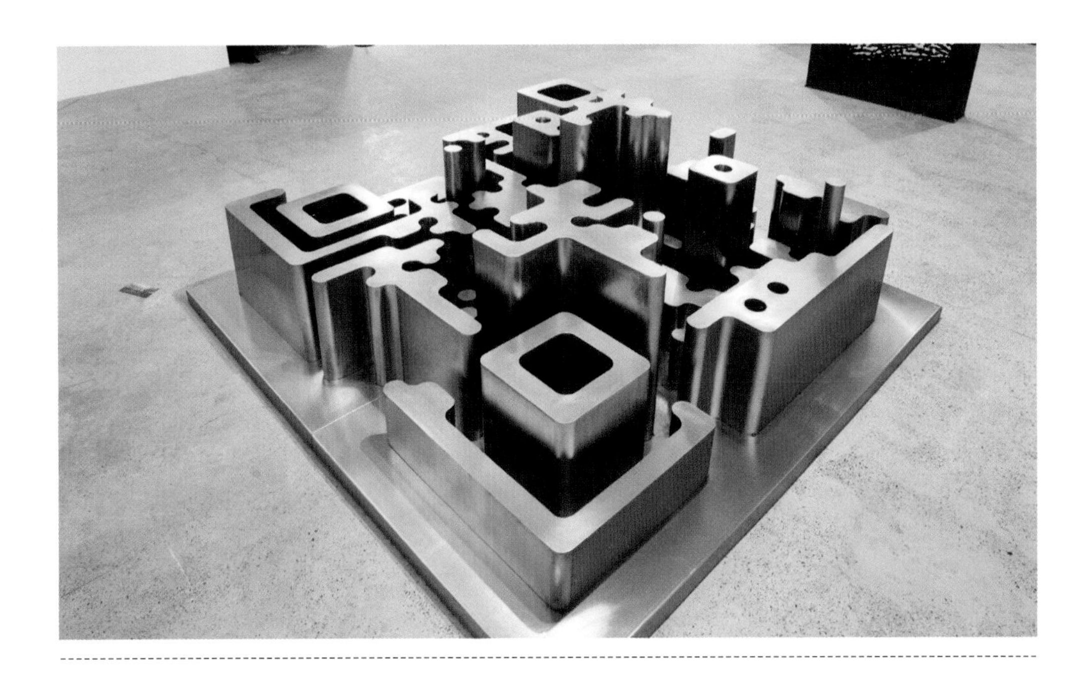

**类型 | Type: 装置 | Installation**

**创作年份 | Year: 2015**

**尺寸 | Size: 215cm × 215cm × 90cm**

**媒材 / 技术 | Mediums/Technology: 不锈钢 | Stainless steel**

**当今社会已进入信息化时代，高科技的发展使二维码的使用在各领域迅速普及。作品取材个人二维码标识，采用二维码图型的基本元素，在原型平面的基础上经过组合调整成立体状，高低错落，犹如现代化建筑鳞栉次比，更如未来世界人类生存的微型城堡。**

**Today we have entered the information age, high-tech makes the usage of 2D barcodes in various fields very fast. This work is made by drawing from personal 2D barcodes identification and the basic elements, then making them from plane to three-dimensional. It was made somewhere high and somewhere low, looks like modern architecture scales, and more like a micro-shaped castle of future human.**

**许旭兵 | Xu Xubing**

1988 年毕业于江南大学工业设计学院，1996 年去日本留学后就读文化服装学园大学，取得硕士学位。现任东华大学服装学院副教授。设计作品曾多次在国内外获奖，举办个人作品发布会等。 Xu graduated from Institute of Industrial Design, Jiangnan University in 1988. He went to Japan in 1996 and was later awarded the Master degree from Bunka Fashion College. He is the Associate Professor of the Donghua University, Fashion and Art Design Institute.

# E-19 印・迹 IMPRESSION

类型 | Type: 雕塑 | Sculpture

创作年份 | Year：2015

尺寸 | Size: 200 cm × 120cm × 25cm × 4

媒材 / 技术 | Mediums/Technology: 钢板 | Steel sheet

**指纹是代表个体的独一无二的核心符号，在中国的语境里，通过指纹，能看到一个人的命运和世界。放在后现代的语境里，指纹也是一个人身份的确认，因此指纹必然负载着社会性和文化性。指纹和契约即是人和物的关系，社会化的关系。**

**Fingerprint is the very core that represents a unique symbol of the individual, in the Chinese context, through the fingerprint, you can see a person's fate and the world; confirmed in a modern context, the fingerprint is a person's identity. Therefore a fingerprint is necessarily loaded with social and cultural nature. The relationship between fingerprints and contracts indicates a social relationship between people and things.**

**严逢林（鞑仁）| Yan Fenglin(Da Ren)**

生于江苏宜兴，现居上海。先后就读于苏州工艺美术学校、天津美术学院、中央美术学院和美国国际科技大学。 Yan was born in Yixing, Jiangsu province. He now lives in Shanghai. He used to study in Suzhou Art and Design Technology Institute, Tianjin Academy of Fine Arts, Central Academy of Fine Art, and International Technological University.

Liang Meiping

# E-20

## 城市编年史
## CITY CHRONICLES

**类型 | Type: 装置 | Installation**

**创作年份 | Year: 2015**

**尺寸 | Size: 15m**

**媒材 / 技术 | Mediums/Technology: 木板、书 | Wood, Book**

**上海，一个被公认为全球最繁忙的城市，竞争以空间新的建设为方法，以速度作为手段。经全球化的催化，各种对城市更新的速度欲望已经在世界各地实践。且看现实生活的配件对缩短时间的各样指导无处不在；如此的人生，究竟是面向什么内容和意义？我们活在一个依靠速度来设计日常生活和生命的更新时代，尤其如何以速度来量化城市、休憩的内容和方式，这如何影响了存在的价值观？此项目重新审视城市更新的速度和欲望的深层辩证。**

**Shanghai, one of the well-known bustling cities in the world, speed can be regarded as the substance supporting our life. Under globalization, the desire for speed has been instilled into different parts of the world, you can observe its proliferation from a range of efficiency gadgets and tips. Leading such a life, what sort of content and meaning are we confronting? We are living in an age that relies speed to structure our lives, particularly in a way that it quantifies things, contents and ways of our work and rest, how exactly does this affect our well being and values? This project reviews the depth dialectic of city innovation speed and desires.**

**梁美萍 | Leung Mee-ping**

创作包括装置、混合媒介、多媒体、公共及社会艺术等。现为香港浸会大学视觉艺术院副教授。个人展览有“星蛹”(2014, 香港)及“不要怪责月亮”(2013, 新加坡)。曾获香港年度艺术家奖(2015), 爱尔兰现代美术馆奖(2014, 爱尔兰)及杰出艺术家。 Leung's works include installation, mix-media, multi-media, public art and community art. She is an associate professor of Hong Kong Baptist University, Academy of Visual Arts. Solo exhibitions include: "Star Pupas" (2014, Hong Kong) and "Don't Blame the Moon" (2013, Singapore). In 2015, she was awarded Annual Artist Award (HKADC, HK); in 2014, was awarded IMMA Fellowship Irish Museum of Modern Art (Ireland).

WIN
METRO
Sun
SEX SELLS
POLICE CALLED
IN ROW OVER
BOY'S FOOTBALL
英中時報

# E-21

## 宅城一族
## HOUSING CITY

类型 | Type: 互动装置 | Interactive Installation

创作年份 | Year: 2015

尺寸 | Size: 可变 | Variable

媒材 / 技术 | Mediums/Technology: 各种日用及工业 | All kinds of household and industrial

当今信息产业的高度发达，让当代年轻人足不出户即能享受便利生活。一种源起于日本“御宅”文化，并带有当代流行文化印记的宅族生活模式正在中国社会悄然兴起。作品《宅城一族》建构了一个象征性的“宅”生活空间，犹如悬疑影片的现场，城市地图、物流单、快递包裹、订餐广告等当代生活图示的线索信息营造出网络通讯与物流产业影响下的“宅族”文化现场。循着现场提示的线索路径，即可探寻当代年轻人的生活印记与心灵迷踪。

The information industry is so developed nowadays that the young generation can enjoy the convenience of their lives without going out. A kind of otaku life mode, which derives from Japanese "Otaku" culture and has the contemporary pop culture imprint, is quietly rising in China. This work, a symbolic "Otaku" living space is constructed as a scene of the suspense movie. The "Otaku" culture scene is created by the clues and information from the city maps, logistics sheet, express parcel, ordering advertisements and other contemporary explanatory charts for life under the influence of network communications and logistics industry. Seeking along with the on-site path to their homes, you can explore the young people's lost soul and the imprint for their lives.

**王启凡 | Wang Qifan**

1989 年生于山东，2012 年毕业于中央美术学院版画系，获学士学位；2015 年毕业于中央美术学院版画系，获硕士学位。现工作、生活于北京。 Wang was born in Shandong Province in 1989. He got his bachelor degree (in 2012) and master degree (in 2015) from Print Making Department of Central Academy of Fine Art. He now works and lives in Beijing.

# E-22 作品第十三号 NO.13

**类型 | Type: 装置 | Installation**

**创作年份 | Year:2015**

**尺寸 | Size: 40cm × 60cm × 15cm × 12**

**媒材 / 技术 | Mediums/Technology: 玻璃 | Glass**

**十二个透明玻璃盒，像是某种特别重要的作品陈列在内，而当观众反复仔细观赏十二个盒子内，最终发现盒内竟空无一物。十二代表了时间，玻璃象征了现代与脆弱、内容与建筑无法配对，毕竟会楼去人空，无法延续。而这十二个盒子就是对这一现象的解读的第十三号作品。**

**There seems to have some important things in the twelve glass boxes. However, they are empty. The number of twelve indicate 12 hours. The glass indicates the weakness and emptiness. These twelve boxes become the thirteenth work of the interpretation of this phenomenon.**

**胡项城 | Hu Xiangcheng**

著名艺术家，现居上海。1977 年毕业中国上海戏剧学院舞台美术系。1986 年旅居日本，从事当代艺术、民间工艺学、造型学的研究，90 年代初，胡项城参与了上海双年展的创立。他是 2010 年上海世博会非洲联合馆艺术总监。 Hu is an artist based in Shanghai. He graduated from Stage art department of Shanghai Theatre Academy, since 1986, lived in Japan, engaged in the contemporary art, folk art, modelling science, and in the early 1990s, he participated and founded the Shanghai Biennale. He was, the art director of the African union pavilion, Expo 2010.

喻红 / Yu Hong

# E-23

## 浮光掠影 + 白领
## DIAPHANOUS FIGURES & WHITE COLLAR

作品名称 | Name: 浮光掠影 | Diaphanous Figures

类型 | Type: 绘画 | Painting

创作年份 | Year: 2013

尺寸 | Size: 38cm × 34cm × 5cm, 38cm × 34cm × 5cm

媒材 / 技术 | Mediums/Technology: 玻璃、玻璃颜料 | Glass, glass pigment colour

作品名称 | Name: 白领 | White Collar

类型 | Type: 绘画 | Painting

创作年份 | Year:2006

尺寸 | Size: 600cm × 110cm

媒材 / 技术 | Material/Technique: 纺织颜料, 丝绸 | Fabric dye on satin

作品反映了在城市更新时代的女性生活及内心，这里交织着个人、家庭、社会、历史的种种因素，交织着幻想、希望、感性、现实，成为城市更新中最细腻的触发点。

**The work reflects the women's life and inner heart in the era of the urban regeneration. There are many factors which could be the individual, family, society, history, fantasy, hope, sensibility, reality. These factors become the most delicate trigger point during the urban regeneration.**

**喻红 | Yu Hong**

1966 年出生于中国北京。著名艺术家，现任教于中央美院油画系。 Born in Beijing, China in 1966. Famous artist, professor of Oil Painting Department, Central Academy of Fine Arts.

天然矿泉水
SHENGSHUI VALLEY
NATURAL MINERAL WATER

# E-24 琐碎 TRIVIALISM

类型 | Type: 综合材料 | Multimedia

创作年份 | Year: 2010

尺寸 | Size: 200cm × 400cm

媒材 / 技术 | Mediums/Technology: 综合材料 | Multimedia

**作品将视角锁定日常生活中的小物件与小场景，以此提供城市更新中的“微尺度”描述。**

**The work targets small objects and small scenes in the daily life in order to provide the description of "micro scale" for the urban regeneration .**

**岳敏君 | Yue Minjun**

中国当代艺术的领军人物，国际著名艺术家。1962 年生，目前生活、创作于北京。 The leading figure of Chinese contemporary art, international famous artists. Born in 1962, Yue now lives and works in Beijing.

# E-25 门 DOOR

**类型 | Type: 装置 | Installation**

**创作年份 | Year: 2015**

**尺寸 | Size: 418cm × 580cm × 545cm**

**媒材 / 技术 | Mediums/Technology: 明代门楼 | Gatehouse of Ming Dynasty**

**作品与城市更新中的新旧之对话，阐释了艺术家眼中的城市与生活的多维关联。**

**It is a dialogue between the work and the old things and new things of the urban regeneration. It describes the multi-dimensional association between the urban and life from the eyes of the artist.**

**张洹 | Zhang Huan**

1965 年出生于河南省安阳市，毕业于北京中央美术学院，现工作生活于上海和纽约。2010 年获 " 年度罗博艺术家 " 称号，2014 年获法国荣誉军团勋章。 Zhang was born in Anyang, Henan Province in 1965. He graduated from China Central Academy of Fine Arts in Beijing, now works in Shanghai and New York. He was awarded the title of "artist of the year" from Robb Report in 2010 and the "Legion of Honour" in 2014.

# E-26 山水城市 LANDSCAPE CITY

**类型 | Type: 雕塑 / 装置 | Sculpture/Installation**
**创作年份 | Year: 2012—2015**
**尺寸 | Size: 240cm × 36cm × 45cm**
**媒材 / 技术 | Mediums/Technology: 不锈钢烤漆 | Stainless steel**

**当城市成为山水的反义词时，造成者始终把山水城市的理念拼贴在城市楼盘之中。作者把山水作为镜中装饰物，悬挂在城市醒目的场所，让城市人面对装饰的山水景观如同水中捞月，可是山水最终是城市人的想象。**

**The artist set the landscape in the mirror as a decorative object such like fish for the moon in the water.**

**申凡 | Shen Fan**

1952 年生于上海。毕业于上海工业高等专科学校，居住、工作在上海。 Shen was born in Shanghai in 1952 and graduated from Shanghai Industrial College. He now lives and works in Shanghai.

# E-27 上海气候 SHANGHAI CLIMATE

类型 | Type: 互动装置 | Interactive Installation

创作年份 | Year: 2015

尺寸 | Size: 可变 | Variable

媒材 / 技术 | Mediums/Technology: 钢架、迷彩尼龙网、气象箱 | Steel frame, Camouflage nylon net, Weather box

“气候交易所”是对人性欲望在各种国家主义战略关系中的视觉解读。“气候交易所”近年来在国际社会中出现，国家之间借助它，通过各种利益争斗，将人类器官与细胞所具有的可在大气层下感观“天人关系”的原始天赋权利，转变成为“公平分配”后的合法权利。可以说它是新型的国际政治权利载体，它的出现，使被消费主义推动的资本主义无限喷张的欲望凌驾于贫困的第三世界之上成为“合法”的事实，从而令被污染的海洋湖泊和陆地上的各种物种都不得不被改变命运，而这种令人惊心动魄的社会现实以前只在世界各大文明的神话传说中出现过。

**"Climate exchange" is the visual interpretation of human nature desire in various national strategic relations.**

**王迈 | Wang Mai**

1972 年生于黑龙江省伊春市，被广泛认为是具有综合才华的艺术家。他自上世纪 90 年代初开始建立其整体艺术的创作构架，从绘画、诗歌到行为表演及装置雕塑，他将各门类艺术形式融会贯通，在思想维度上立足于国际视野，而其跨媒介的表现语言则根植于深厚的传统文脉基因。 Born in Yichun, Heilongjiang Province in 1972, Wang Mai is widely regarded as a multi-media pioneer, and one of the most versatile artists of his generation. Since the 1990s, his work has spanned from painting and poetry to performance and installation, deconstructing social ideologies to reflect on global issues.

上海市文化广播影视管理局

# E-28

## 云海间
## CLOUDS

类型 | Type: 雕塑 | Sculpture

创作年份 | Year: 2015

尺寸 | Size: 7m × 3.85m × 1.46m

媒材 / 技术 | Mediums/Technology: 亚克力板 | Acrylic plate

此方案选取山水画为原型，用开阔的胸怀、深细敏感的审美嗅觉、不同材质不同手法的表现方式来描绘山水画的优美壮丽，标示着对于社会和谐静溢谧的憧憬及希翼，同时阐述了在这一大时代背景下诞生的上海中心对中国传统文化艺术的尊重以及传承的决心，并以新的视角对传统文化艺术进行新的解读和诠释。体现了上海中心作为当代最高文明成就的集合体，将激励并推进现代与未来的文明交流与成就创造。上海中心本身也是意念的“山水”。

**This work selects the landscape painting as the prototype, which embodies the Shanghai Centre as a collection of the highest achievements of modern civilization, will encourage and promote the modern and the future of cultural communication and achievements. Shanghai Centre itself is also a "landscape of idea".**

**米丘 | Mi Qiu**

跨领域艺术家，致力于推动艺术与科技结合的人类文明母题。米丘的大量作品，特别是他的行为、装置和大型环境作品，都显示出跨世界、跨领域的对话意识。A cross-disciplinary artist committed to promoting the integration of art and technology, an undertaking reflecting human civilization.

总策展人：伍江
开幕日| Opening
展期| Duration：2015
上海 徐汇 龙腾大道
2555 Longteng
徐汇区人民政府

# E-29

## 问道——孔子问道老子
## DIALOGUE — CONFUCIUS ASKED LAOZI ABOUT TAO

类型 | Type: 雕塑 | Sculpture

创作年份 | Year: 2012

尺寸 | Size: 孔子 | Confucius: 7.7(h) × 1.81(w) × 1.80(t)m; 老子 | Laozi:7.86(h) × 1.85(w) × 1.82(t)m

媒材 / 技术 | Mediums/Technology: 青铜 | Bronze

《史记 · 老子韩非列传》中记载，孔子适周，曾问道于老子。老子时任周之守藏室史，新沐披发，与孔子语以深藏若虚、逢时而动的思想观念。孔子归，以告弟子：龙乘风云而上天，非其所能知也，今见老子，其犹如龙也！

**Upon arriving in the Capital of Zhou Dynasty, Confucius paid a special visit to Laozi, the Keeper of the Imperial Archives and asked him about Tao. Laozi just had a bath and his hair hung over his shoulders, and he shared with Confucius the concept that a wise man rises when his time comes; but if it does not, he keeps his treasures buried deeply to make it look as if he had none. Confucius returned and told his students - I cannot tell how the dragon mounts on the wind through the clouds, and rises to heaven. Today I have seen Laozi, and can only compare him to the dragon.**

**--Excerpt from *The Story of Laozi and Han Feizi, the Grand Historian***

**吴为山 | Wu Weishan**

全国政协委员，中国美术馆馆长，中国美术家协会副主席，中国雕塑院院长，教授、博士生导师。Wu serves as a Member of the National Committee of the Chinese People's Political Consultative Conference (CPPCC), President of the National Art Museum of China, Vice President of Chinese Artists Association. He is also a professor and PhD supervisor.

# E-30 不能停 NO STOP

类型 | Type: 装置 | Installation

创作年份 | Year: 2015

尺寸 | Size: 15m × 10m

媒材 / 技术 | Mediums/Technology: 混凝土搅拌机等 | Concrete mixing machine and etc.

混凝土搅拌罐是建筑业不可或缺的重要机械。它一旦启动就必须不停旋转，否则就会导致混凝土凝固，半途而废。城市的更新与再造是中国现代化进程和经济社会运转的一种体现，这一进程同样是只要启动就不能停止……吐出来又吐进去的循环过程，体现了事物发展的客观规律：吐故纳新。任何事物的发展都需要一个复杂的过程，在循环中吸收、在温故中消化、在变化中发展、在反复深化，在痛苦中创新……循环的体验即是产生和创造新事物的必经之途。更新的过程也是消化过程，在结合和发展中会出现各种反应，甚至出现虚幻的泡沫。由于蕴含着巨大的能量，因此，事物一定会克服困难，向前发展。

Concrete mixing tank is an indispensable construction machine. It must be constantly rotating once started, otherwiseconcrete will solidify. The updating and rebuilding of the city is a reflection of China's modernization process and social economic operation. This process is the same: Once started, it will never end... This circulation process of spit out and sucked in shows the objective law of things development: get rid of the stale and take in the fresh. The development of everything needs to be a complex process, absorb in the cycle, digest in the review develop in the change, deepen again and again, innovate in pain... The experience of circulation is the only way to produce and create new things. The update process is also the process of digestion, and in the process of combination and development will appear various reactions, even illusory bubbles. Because of the vast amounts of energy, therefore, things will overcome the difficulties, forge ahead!

**马堡中 | Ma Baozhong**

1965 年出生于黑龙江省，1987 年考入中央美术学院油画系。职业艺术家，现工作、生活在北京。 Born in Heilongjiang province in 1965, Ma started his study in Oil painting department of Central Academy of Fine Art in 1987. He is now a professional artist, works and lives in Beijing.

西岸艺术中心 WEST BUND ART CENTER
龙腾大道2555号 | 2555Long Teng Avenue

# E-31 无语空间 SILENT SPACE

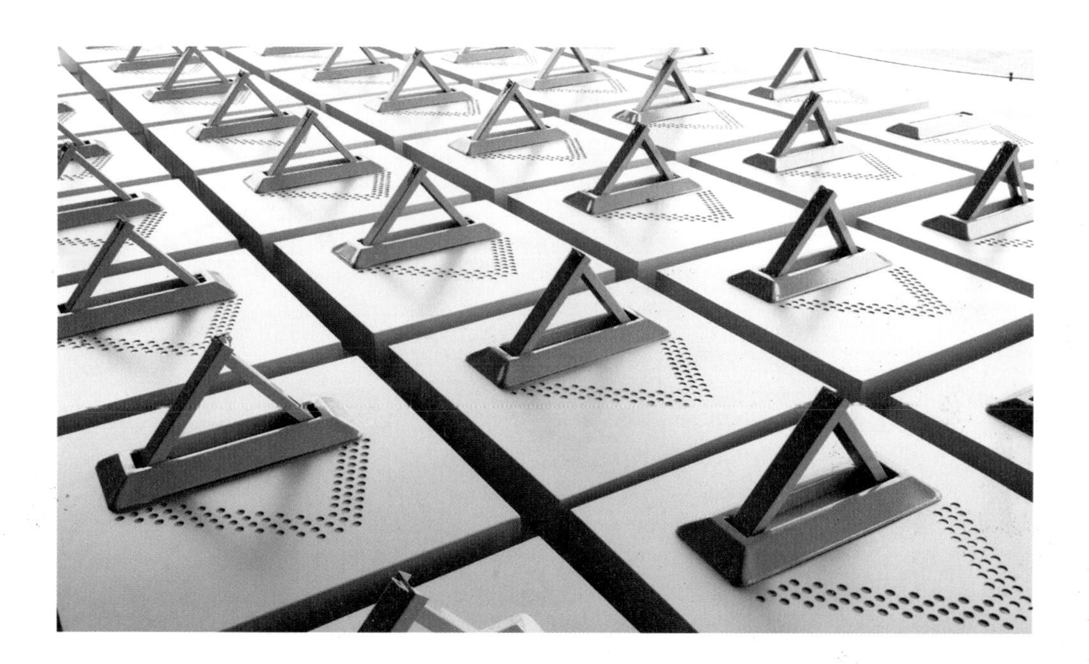

**类型 | Type: 互动装置 | Interactive Installation**

**创作年份 | Year: 2015**

**尺寸 | Size: 80cm × 80cm × 50cm × 64**

**城市更新带来了交通方式的更新，汽车取代了自行车，同时有限的城市空间被变成了停车场，占据了人群的正常行走。作品以 100 个左右汽车地锁放置在展览馆的广场。地锁上装有发光器，在夜晚能让参观者感受到大自然原生态的绚烂，以此来揭示城市更新中如何尊重大自然生态与人文生活的美好。**

**The urban regeneration has changed the mode of transportation. The car replaced the bike. More and more space in the city has been used as parking space. 100 ground locks with light device are settled on the square of the exhibition hall. People could experience the gorgeous of the nature in the night. In the same time, they should think about the balance between the urban life and nature.**

**孙哲政 | Sun Zhezheng**

1963 年出生于南京，1982 年考入苏州工艺美术学院，1987 年毕业于南京师范大学美术学院油画研修班。现居南京，职业艺 术 家。 Sun was born in Nanjing in 1963. He studied in Suzhou Art and Design Technology Institute in 1982. In 1987, he graduated from Academy of Fine Arts, Nanjing Normal University. He now lives in Nanjing and works as a professional artist.

CHINA

# E-32 中国制造 MADE IN CHINA

类型 | Type: 装置 | Installation

创作年份 | Year: 2005

尺寸 | Size: 1200cm × 240cm × 265cm

媒材 / 技术 | Mediums/Technology: 考顿钢 | Corten steel

在过去 20 年里，隋建国始终在对逝去的文明进行反思，力图在有限的三维空间中完成无限的观念实践。隋建国在作品中呈现出一个又一个表征，却又从未陷入某个文化符号的桎梏，始终以精神性的观念主导一切，突破自我。通过此次展出的一系列作品，我们可以清楚地看到艺术家内心思考的心路历程。

Over the past 20 years, Sui Jianguo has continuously reflected upon civilizations of the past, attempting to realize his limitless conceptual practice within a limited three-dimensional space. In his works, Sui Jianguo presents symbol after symbol, yet manages to never be constrained by the "shackles" of any one cultural symbolism in particular. All the while, he makes use of his concept of spirituality as a guide, breaking away from the self. Through the works displayed in this exhibition, we can clearly see the artist`s internal thought process.

**隋建国 | Sui Jianguo**

1956 年生。著名艺术家，现为中央美术学院雕塑系系主任。 Born in 1956, the artist now is the Director of Sculpture Department, China Central Academy of Fine Arts.

# E-33 如意 RU YI

类型 | Type: 雕塑 | Sculpture

创作年份 | Year: 2011

尺寸 | Size: 645cm × 255cm × 55cm; 485cm × 210cm × 70cm; 185cm × 200cm × 62cm

媒材 / 技术 | Mediums/Technology: 不锈钢 | Stainless steel

公共雕塑《十示如意》的形象来源于中国传统的如意、云纹、玄机等表意符号，通过对符号表意特征的研究以及艺术家个人化“十”字符号语言的嵌入，反映出中国传统文化图形元素的源流以及当代语境中新的转化、延伸、发展、更新的可能性。对传统符号重新注释的过程，本身包含着新的角度的再现逻辑；当象征符号被放大、被变异、被重新创造，以往的经验和历史线索可能被阻断；当传统中这些微小器型一旦被放大至几十倍时，文化记忆和意义指向都会面临更多的挑战。

**The image of public sculpture *Appearance of Crosses - Ru Yi*, comes from the Chinese traditional ideographs like Ru Yi — an S-shaped ornamental object, usually made of jade, formerly a symbol of good luck' cloud pattern, Xuan Ji and etc., by researching the meaning of the symbols and inserting of the artist's personalized Cross, the source of Chinese traditional pattern elements and, of which, the possibility of the transform, expedition, development and renovation have been revealed in the contemporary context. The process of the re-annotation of the traditional symbols contains the reappearance of the logic from a new angle. When the symbols are magnified, changed, and recreated, the experience and the historic clues might be interdicted, and when these micro shapes are enlarged for dozens of times, the culture memory and significance direction would face more challenges.**

**丁乙 | Ding Yi**

1962 年生于上海，现工作、生活于上海。1983 年毕业于上海工艺美术学校，1990 年毕业于上海大学美术学院。丁乙的创作领域包括绘画、雕塑、空间装置和建筑。“十”字以及变体的“X”是他主要的视觉符号，用以超越和对抗当时中国典型的政治、社会寓言绘画。从 80 年代后期开始，他将这一标志作为结构和理性的代表，以及反映事物本质的图像表现的代名词。 Ding Yi (b. 1962) currently resides in Shanghai. He graduated from Shanghai Arts & Crafts Institute in 1983 and later from Academy of Arts, Shanghai University with B.F.A. in 1990. The practice of Ding Yi encompasses painting, sculpture, spatial installation and architecture. He works primarily with "+" and its variant "x" as formal visual signals, above and against the political and social allegories typical of painting in China. He chose this sign in the second half of the 80s as a synonym.

# E-34 从黄浦江出发 STARTING FROM THE HUANGPU RIVER

**类型 | Type: 雕塑 / 装置 | Sculpture/ Installation**

**创作年份 | Year: 2015**

**尺寸 | Size: 2580mm × 15000mm × 10mm**

**媒材 / 技术 | Mediums/Technology: 铁板 | Iron plate**

**以黄浦江的造型为例，作品以镂空钢板作为基本造型，用特种工艺篆刻“海上丝绸之路”的城市、符号、图案等图文内容。透过丝绸之路参看上海城市的繁荣，通过上海城市参看丝绸之路的活力，透视现代城市的文化更新。**

**The work shows the city, symbols, patterns and the other graphic content of "Marine Silk Road" by using the special craft carving. We shall see the prosperity of the Shanghai city by review the "Silk Road". We also can see the vitality of the "Silk Road" through the city of Shanghai. It is the cultural renewal of the modern city.**

**NC 小组 | Group NC**

NC 小组组建于 2015 年，致力为跨文化、跨学科、跨媒体艺术提供创作，通过学术跨界对话将艺术的新锐探索转化为大众公共教育和资源。2015 年参与第一届亚洲双年展。 The group was established in 2015, dedicated to artistic work in the area of cross culture, cross disciplinary, cross media. In 2015, the group took part in the first Asian biennial.

# 特展：越 · 上海

# SPECIAL EXHIBITION: YUE · SHANGHAI

F

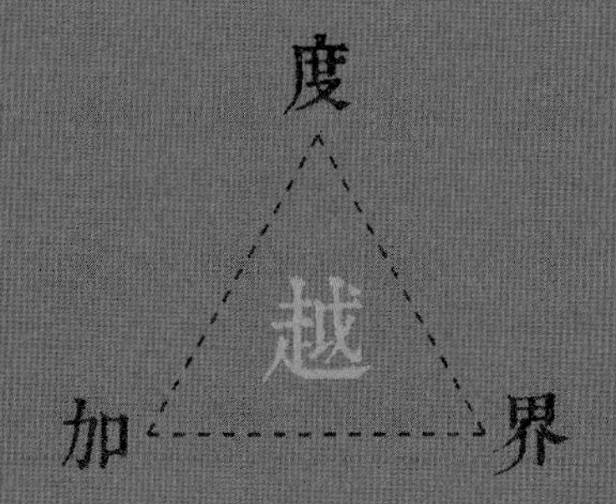
度
越
加
界

越 | 上海
特展·越上海
YUE | SHANGHAI

## 越・上海

章明 & 张姿，
奚文沁，
王林

在中国城市发展的语境中，上海的城市更新以先行者的角色引领并推动了中国城市更新的当代进程。

在世界城市发展的语境中，上海的城市更新因其独特性与多元性的面貌在全球城市更新的浪潮中独树一帜。

越，在中文中含义颇多，也充满了不确定性。

越度——表达对身体的移动性与关联体验的关注；

越界——表达对感知的边界性与边界消解的关注；

越加——表达对发展的过程性与特质彰显的关注。

这里呈现的上海，不是一个孤立的地理意义上的存在，而是类似于新陈代谢式的更替与成长的绵长流程，是在庞杂城市背景下的去除与填补间的平衡。它是一个鼓励身体的移动与关联体验的城市。

这里呈现的上海，不是一个条块分明的功能意义上的存在，而是一个不断融合与分离的可变体系，是在共性与差异性间谋求共同的生存环境。它是一个允许杂陈共处的城市。

这里呈现的上海，不是一个片断的记忆再现式的存在，而是由人文情怀与价值认同不断积累的丰厚积层。它是一个充满浓厚的地方特质并乐于彰显这种特质的城市。

这也是"越•上海"置身于"2015上海城市空间艺术季"的理由和出发点。展览共征集三十余年来上海城市更新的典型性案例作品五十余件，用全景的、地理索引的方式呈现出一个没有主从结构、没有先后位序的公平、并置、开放的展陈空间。

上海的城市更新并不能跳脱中国城市发展的语境而存在，它始终置身于中国上世纪80年代开始的快速城市化/现代化进程之中。同时，上海由于其独特的条件与地位，注定在强大的权力、资本以及全球化力量所裹挟的消费主义浪潮中所发生的社会与城市空间的变迁更为迅猛、剧烈，所催生出的案例与经验更加丰富多元，现象背后的价值认同也更为复杂与微妙。因此，回溯上海城市更新的当代进程、梳理城市更新的理念是一件紧迫且有意义的事。

## YUE · SHANGHAI

**Zhang Ming & Zhang Zi,**
**Xi Wenqin,**
**Wang Lin**

In the context of Chinese urban development, Shanghai promotes the contemporary Chinese urban regeneration process as a pioneer.

In the context of the world urban development, Shanghai is unique in the global wave of urban regeneration for its unique features and diversity.

"Yue", a Chinese character, contains multiple connotations and embraces uncertainties.

It could mean cross-dimension, referring to the focus on the mobility and related experience of human beings.

It could mean cross-boundary, which should be understood as the focus on the perceivable boundary and the dissolution of such a boundary.

It also has the meaning of"more", which focuses on the course of the development and the characteristics of certain subjects.

The concept of Shanghai here is not a merely geographical idea, but more like a concept of vicissitudes and the gradual and long-term course of development. The concept of Shanghai here refers to something that fills out the vacancy between the elimination and creation in this city. It refers to a city that encourages the mobility of humans and their experience.

The concept of Shanghai presented here is not a city with clearly separated functions, but more like a changeable system that remains in variation and a living environment in which generality coexists with otherness. It is a city that tolerates differences.

The concept of Shanghai is not the reproduction of a piece of memory, but more like the outcome of the growth of humanity and the acceptance of universal values. It is a city that features and enjoys featuring its local characteristics.

This is the fundamental reason and idea to have the "Yue · Shanghai" in the Shanghai Urban Space Art Season 2015. This exhibition totally collected more than 50 typical cases about the variations of Shanghai city in the recent 3 decades and created an exhibition site that did not present exhibited items in any given order due to its emphasis on equality.

The urban regeneration process of Shanghai could not be fully interpreted without a proper understanding on the background of the rapid urban development in China. The rapid growth of Shanghai was made possible under the urbanization and modernization of China since 1980s. The unique advantages and the special position of Shanghai in China determined its ability to have the strong and rapid urban development momentum in the wave of consumerism driven by political power, capital power and the strength of globalization. Under this situation, the cases and experiences about the development of Shanghai were getting more diverse while the universally accepted values expressed in these cases became much more complicated. Thus, it is urgent and significant to retrospect the course of urban regeneration of Shanghai and clearly interpret the concept of urban regeneration.

四行仓库

## 策展理念

越度——表达对身体的移动性与关联体验的关注。
越界——表达对感知的边界性与边界消解的关注。
越加——表达对发展的过程性与特质彰显的关注。

以历史演进为线索，系统性地呈现上海城市更新的背景、理念及发展历程；以全球化为背景，研讨上海城市更新的模式对未来的意义；以在地性为框架，用地理图像为展示基础的展陈形式，全景式地展示上海城市更新的典型性案例。

## 展览结构

突破传统展览布局模式，将“上海特展”与二层“艺术介入公共空间”展、一层公共接待空间及公共交通空间融通为一体，相互穿插渗透。用一个全景的、地理索引的方式呈现出一个没有主从结构、没有先后位序的公平、并置、开放的展陈效果。整个展区以红色水管为母题组成的放射状景框，形成一个整体性的展陈空间，景框与景框之间又生成了独立展陈单位。巨幅上海地图于二层展览区域内的地面满铺，以直接清晰的方式向参观者展示上海城市更新的典型性案例。

## CURATORIAL CONCEPT

Cross-dimension, referring to the focus on the mobility and related experience of human beings.

Cross-boundary, which should be understood as the focus on the perceivable boundary and the dissolution of such a boundary.

"More", which focuses on the course of the development and the characteristics of certain subjects.

Taking the historical evolution as a clue, exhibiting Shanghai urban regeneration about its background, philosophy and development process systematically; Taking the globalization as the background, discussing Shanghai urban renewal model for the future significance; Taking the In-Situ as a framework, displaying Shanghai Urban Regeneration typical cases panoramically with geography-based form of exhibition.

## EXHIBITION STRUCTURE

Breaking the traditional exhibition layout mode, mixing the space of "Shanghai Special Exhibition", "Art Intervention in Public Space" exhibition on the 2nd floor, public reception space and public circulation space on the 1st floor together. Presenting an equal, parallel, open-ended, no master-slave exhibition structure with a panoramic and geography-based mode.

The pavilion is a holistic exhibition space, was formed of radial rectangle frame consisting of red pipes and the frame can also generate a series of independent exhibition units. A huge Shanghai map was posted on the ground floor, covered the whole exhibition area, in order to show Shanghai Urban Regeneration typical cases to visitors directly and clearly.

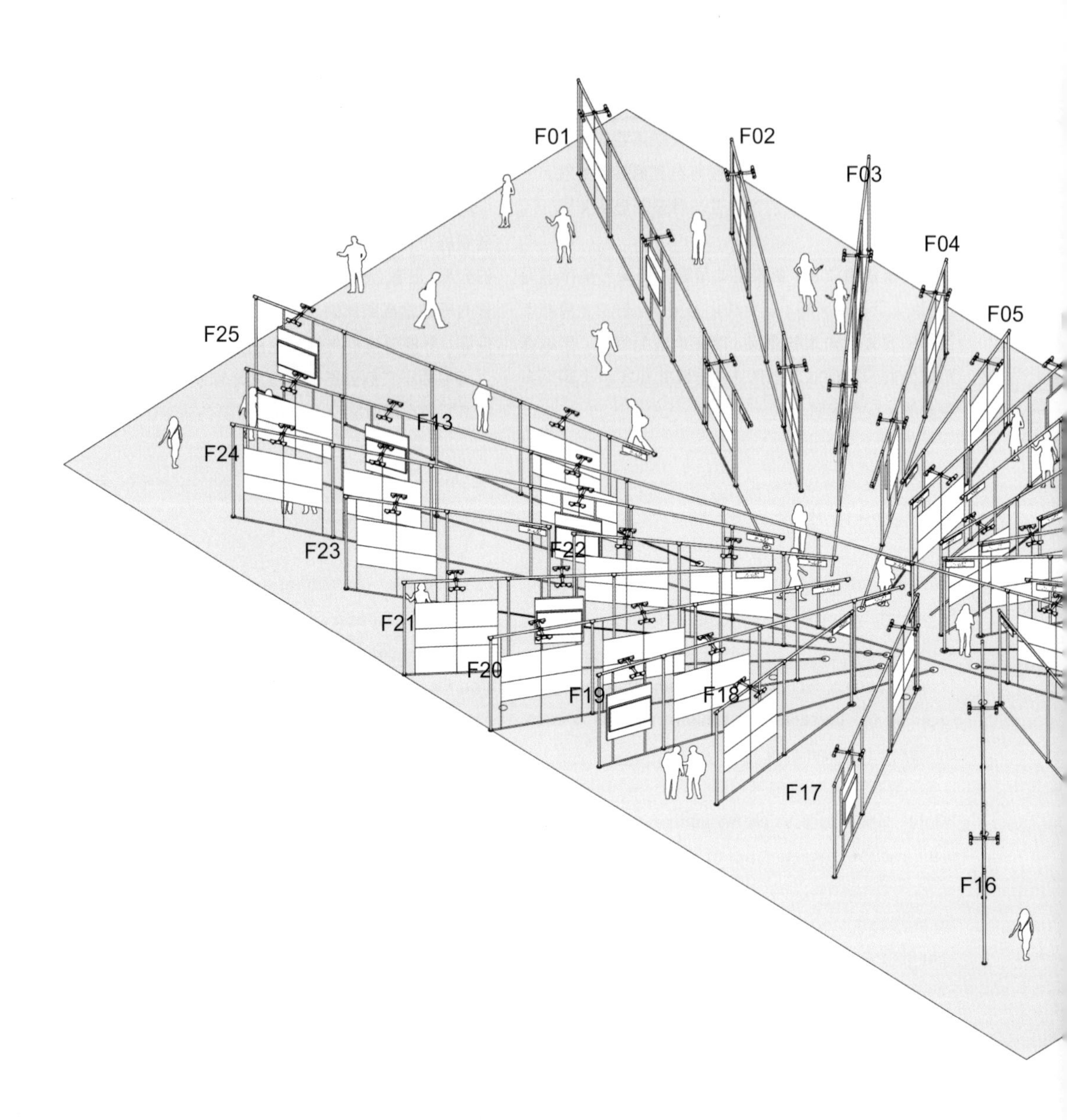
F01
F02
F03
F04
F05
F25
F13
F24
F23
F22
F21
F20
F19
F18
F17
F16

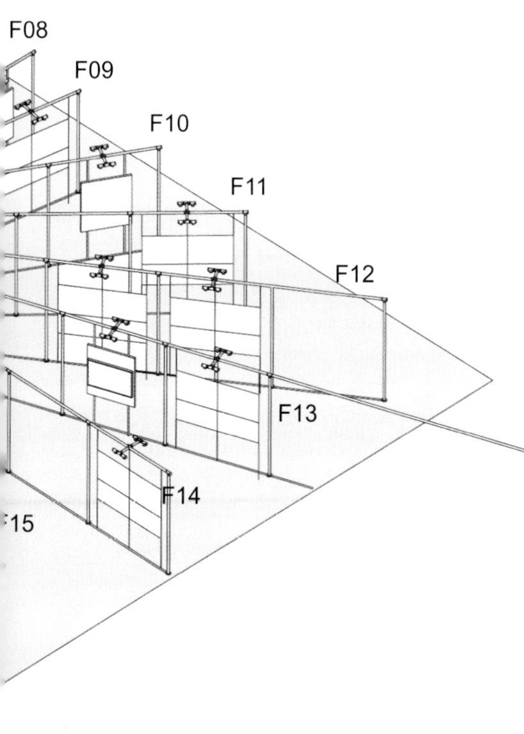


F01：枫泾古镇、东安路（中山南二路——龙华中路）城市空间提升项目、徐家汇街道社区服务设施及周边环境整治项目；
F02：月湖雕塑公园、区县展活动内容展板；
F03：上海城市雕塑艺术中心、朱家角淀浦河南岸尚都里；
F04：徐汇风貌区道路保护规划及整治实施计划研究、区县展活动内容展板；
F05：区县展活动内容展板；
F06：愚园路历史文化风貌街区；
F07：静安雕塑公园—上海自然博物馆、曹杨新村；
F08：同孚绿地、桃浦科技智慧生态城；
F09：上海飞联纺织厂厂房改扩建工程；
F10：苏州河滨河地区开发；
F11：东斯文里、区县展活动内容展板；
F12：四行仓库、环上海大学国际影视产业园区、炮台湾湿地森林公园；
F13：黄浦江两岸综合开发、上海历史文化风貌区保护规划、宝山钢雕公园、浦南运河 & 江海经济园区；
F14：新江湾城、创智天地；
F15：多伦路风貌保护道路、上海船厂地区；
F16：上海音乐谷公共艺术区、上海国际时尚中心、序言展板；
F17：外滩滨水区、外滩源；
F18：老码头、区县展活动内容展板；
F19：南京东路步行街、区县展活动内容展板；
F20：上海世博会地区后续利用、太平桥地区；
F21：世博会城市最佳实践区；
F22：八号桥、延中绿地；
F23：召稼楼古镇（革新村）；
F24：思南公馆、田子坊；
F25：廊下新农村建设、区县展活动内容展板、序言展板。

F01: Fengjing Ancient Town, Renovation of Urban Space in Dongan Road, Renovation of Community Service Facilities;
F02: The Moon Lake Sculpture Park, The District Exhibition Activities Board;
F03: Shanghai City Steel Sculpture Park, Shangduli, Southern Bank of Dianpu River in Zhujiajiao;
F04: Xu-Hui Historical Streets: Detailed Conservation Planning & Renovation Action Plan Study, The District Exhibition Activities Board;
F05: The District Exhibition Activities Board;
F06: Yuyuan Road Historical and Cultural Site;
F07: Jing'an Sculpture Park- Shanghai Natural History Museum, Caoyang New Village;
F08: Tongfu Greenbelt, Taopu Smart Eco-City of Science and Technology Development;
F09: Shanghai Feilian Textile Mill Plant Expansion Project;
F10: Development of Suzhou River Waterfront;
F11: Dong Siwen Li, The District Exhibition Activities Board;
F12: Sihang Warehouse, International Film and Television Industry Park along Shanghai University, Paotai Wan Wetland Forest Park;
F13: Comprehensive Development of Huangpu River Waterfront, Conservation Planning for Historic Conservation Areas in Shanghai, Baoshan Steel Sculpture Park, Punan Canal & Jianghai Industrial Zone;
F14: New Jiangwan City, Knowledge and Innovation Community(KIC);
F15: Duolun Road Conservation Pathway, Shanghai Shipyard Area;
F16: Shanghai Music Valley Public Art Zone, Shanghai International Fashion Center, Foreword Board;
F17: Waterfront Area of the Bund, "Bund Origin" Area;
F18: The Old Pier, The District Exhibition Activities Board;
F19: Nanjing Road (East) Pedestrian Street, The District Exhibition Activities Board;
F20: Shanghai Post-Expo Development, Taipingqiao Area;
F21: The Urban Best Practices Area in the Expo;
F22: Bridge 8, Yanzhong Green Space;
F23: ZhaoJialou Ancient Town (Gexin Village);
F24: Sinan Mansion, Tianzi Fang;
F25: Construction of new countryside in Langxia, The District Exhibition Activities Board, Foreword Board.

**参展机构 | Participating Institution**

上海市浦东新区人民政府、黄浦区人民政府、静安区人民政府、徐汇区人民政府、长宁区人民政府、普陀区人民政府、闸北区人民政府、虹口区人民政府、杨浦区人民政府、宝山区人民政府、闵行区人民政府、嘉定区人民政府、金山区人民政府、松江区人民政府、青浦区人民政府、奉贤区人民政府、崇明县人民政府 / People's Government of Shanghai Pudong New District, Huangpu District, Jing'an District, Xuhui District, Changning District, Putuo District, Zhabei District, Hongkou District, Yangpu District, Baoshan District, Minhang District, Jiading District, Jinshan District, Songjiang District, Qingpu District, Fengxian District, Chongming County.

**参展作品**

**跨区：** 黄浦江两岸综合开发、苏州河滨河地区开发、上海历史文化风貌区保护规划 / **浦东新区：** 上海世博会地区后续利用、上海船厂地区 / **黄浦区：** 外滩滨水区、外滩源、老码头、八号桥、思南公馆、田子坊、太平桥地区、南京东路步行街、世博会城市最佳实践区、延中绿地 / **静安区:** 静安雕塑公园—上海自然博物馆、东斯文里、同孚绿地 / **徐汇区:** 徐汇风貌区道路保护规划及整治实施计划研究、东安路（中山南二路—龙华中路）城市空间提升项目、徐家汇街道社区服务设施及周边环境整治项目 / **长宁区：** 愚园路历史文化风貌街区、上海城市雕塑艺术中心 / **普陀区：** 曹杨新村、桃浦科技智慧生态城 / **闸北区：** 环上海大学国际影视产业园区、四行仓库 / **虹口区：** 上海音乐谷公共艺术区、多伦路风貌保护道路 / **杨浦区：** 新江湾城、创智天地、上海国际时尚中心 / **宝山区:** 宝山钢雕公园、炮台湾湿地森林公园 / **闵行区:** 召稼楼古镇（革新村） / **嘉定区:** 上海飞联纺织厂厂房改扩建工程 / **金山区:** 廊下新农村建设、枫泾古镇 / **松江区:** 月湖雕塑公园 / **青浦区:** 朱家角淀浦河南岸尚都里 / **奉贤区:** 江海经济园区 / **崇明县:** 大地景观艺术计划

**Cross-District Projects:** Comprehensive Development of Huangpu River Waterfront, Development of Suzhou River Waterfront, Conservation Planning for Historic Conservation Areas in Shanghai / **Pudong New District:** Shanghai Post-Expo Development，Shanghai Shipyard Area / **Huangpu District:** Waterfront Area of the Bund, "Bund Origin" Area, the Old Pier, Bridge 8, Sinan Mansion, TianZi Fang, Taipingqiao Area, Nanjing Road (East) Pedestrian Street, The Urban Best Practices Area in the Expo, Yanzhong Green Space / **Jing'an District:** Jing'an Sculpture Park- Shanghai Natural History Museum, Dong Si Wen Li, Tongfu Greenbelt / **Xuhui District:** Xu-Hui Historical Streets: Detailed Conservation Planning & Renovation Action Plan Study, Renovation of Urban Space in Dongan Road, Renovation of Community Service Facilities / **Changning District.** Yuyuan Road Historical and Cultural Site, Shanghai City Steel Sculpture Park / **Putuo District:** Caoyang New Village, Taopu Smart Eco-City of Science and Technology Development / **Zhabei District.** International Film and Television Industry Park along Shanghai University, Sihang Warehouse / **Hongkou District.** Shanghai Music Valley Public Art Zone, Duo Lun Road Conservation Pathway / **Yangpu District.** New Jiangwan City, Knowledge and Innovation Community(KIC), Shanghai International Fashion Center / **Baoshan District:** Baoshan Steel Sculpture Park, Paotai Wan Wetland Forest Park / **Minhang District:** ZhaoJialou Ancient Town (Gexin Village) / **Jiading District:** Shanghai Feilian Textile Mill Plant Expansion Project / **Jinshan District:** Construction of new countryside in Langxia, Fengjing Ancient Town / **Songjiang District:** The Moon Lake Sculpture Park / **Qingpu District:** Shangduli, Southern Bank of Dianpu River in Zhujiajiao / **Fengxian District:** Jianghai Industrial Zone / **Chongming County:** Chongming Land Art Plan

## 8 号桥 | BRIDGE 8

上海特色创意园区 8 号桥位于建国中路 8-10 号，占地 7 000 多平方米，总建筑面积 12 000 平方米，曾是旧属法租界的一片旧厂房，解放后成为上海汽车制动器厂旧厂房。

2004 年初，一次偶然的机遇引发了 8 号桥的打造开发，同时也开启了旧厂房改造更新利用这一全新的转型方式和运营模式。经过新的设计与改造后注入时尚、创意的元素，8 号桥成为了沪上时尚创意园区之一。

Shanghai Bridge 8 Creative Industry Parksituated at No. 8-10, Middle Jiangguo Road. It occupies an area of 7,000 $m^2$ and has a total floor area of 12,000 $m^2$. Used to be an old factory of former French Concession, and after the foundation of the P. R. C, it was converted to the factory of Shanghai Automobile Brake Co., Ltd.

In early 2004, the transformation of Bridge 8 was launched through an occasional opportunity, which led to the utilization of new renovation and operation mode in the transformation and upgrade of the old factory in downtown Shanghai. New design poured fashionable and creative elements into the tradtional area and turned it into one of the most fashionable creative industry parks in Shanghai.

## 创智天地 | KNOWLEDGE INNOVATION COMMUNITY (KIC)

创智天地位于上海杨浦区五角场城市副中心，项目规划用地面积 49 公顷，总建筑面积 76 万平方米。创智天地秉持校区、社区、科技园区“三区融合，联动发展”的理念，复旦、同济、财大等十余所知名高校和百余家科研院所环绕其旁。

创智天地以独具特色的大学路为主轴，将江湾体育场、创智天地广场 / 企业中心和创智坊连为一体。该地区土地利用高度混合，创造更多智慧交流的空间和机会，从而形成充满活力的生活、娱乐、工作、学习多功能融合的综合社区。社区由许多小街区组成，街区之间由尺度宜人的街道和广场相连，同时布置较多小型广场和公共空间供社区共享。全区鼓励步行和骑车，地铁和公交站点方便可达，另外还设有公共自行车站，方便所有游客和居民在日常生活中步行或骑车出行。

Located at Wujiaochang Sub-center of Shanghai Yangpu District, KIC has a planning area of 49 hectares, with a total construction area of 760,000 $m^2$. KIC upholds the idea of "Three-Zone Integration for Interlinked Development" as school campuses, communities and scientific and technological parks. KIC is enriched by 10 renowned neighboring universities such as Fudan University, Tongji University, and Shanghai University of Finance and Economics and so on as well as more than one hundred scientific research institutions.

With unique University Avenue as the principal spine, KIC connects the Jiangwan Stadium, KIC Plaza / Corporate Center and KIC Village together. The highly mixed land use creates more public space and opportunities for intellectual exchange, thereby forming a dynamic multi-function integrated community where people live, play, work and learns. The community is made up of many small blocks, which are connected by the pleasant streets and squares with small squares and spaces for public sharing. Walking and cycling are encouraged here, and the metro and bus stations are within easy reach in addition to public bicycle stations, so that all visitors and residents can freely walk or cycle in everyday life.

## 思南公馆 | SINAN MANSION

思南公馆总占地面积约 5 万平方米，是衡山路 - 复兴路历史文化风貌区的重要组成部分，拥有 51 栋各式花园洋房，包括独立式花园洋房、联立式花园洋房、带内院独立式花园洋房、联排式建筑、外廊式建筑、新式里弄、花园里弄、现代公寓等多种建筑样式，是上海近代居住类建筑的集中地。

作为市中心独立成片花园住宅最集中的区域之一，黄浦区将其定位为"具有上海近代独特文化和历史特点的高品质生活居住、商业休闲社区"，保护花园住宅区的传统环境，包括庭院空间、沿街界面、内部通道、空间节点等，保持庭院空间的完整性、住宅的私密性，营造整体宁静、优雅的氛围。通过周围腹地提供支持性的休闲娱乐服务设施，使其整体成为一个重要的城市文化景点。

Occupying an area of 50,000 $m^2$, Sinan Mansion is an important part of the Hengshan Road – Fuxing Road Historic Conservation Area. It contains 51 garden houses, including independent garden houses, row garden houses, independent garden houses with inner courts, row-buildings, veranda-style buildings, new-style lanes, garden lanes, and modern apartments, a complete collection of modern Shanghai residential buildings.

Boasting one of the most centralized regions of garden mansions in downtown Shanghai, Sinan Mansion is defined by the Huangpu District Government as "an upscale residential, commercial, and recreational hub of Shanghai with unique modern cultural and historical features". The traditional environment of garden mansions, including courtyards, interfaces along the street, inner passageways, and space connections, has been conserved, and the completeness of courtyards, the privacy of residences, and a serene and elegant atmosphere was guaranteed and created. Supported by surrounding entertainment and leisure facilities, Sinan Mansion has become an important urban cultural core.

## 太平桥片区 | TAIPINGQIAO AREA

上海太平桥片区毗邻上海市中心主要商业街之一——淮海中路，规划范围东临黄陂南路，南至自忠路，西抵马当路，北邻太仓路，占地面积 51 公顷，规划总建筑面积 125 万平方米。

太平桥地区是“九五”期间老卢湾区重点的旧区改造项目。1996 年地区启动石库门建筑改造项目，改变原先的居住功能，赋予它新的商业、文化与公共活动功能，把百年的石库门旧城区，改造成一片“新天地”，令整片区域成为一个综合功能的国际化社区。

Shanghai Taipingqiao Area is adjacent to Middle Huaihai Road, one of the main commercial streets in the center of Shanghai, with the planning area east to South Huangpi Road, south to Zizhong Road, west to Madang Road, and north to Taicang Road. It covers an area of 51 hectares, and the total building area is 1.25 million square meters.

The Taipingqiao project is one of the key urban transformation projects in old Luwan District during the "Ninth Five-Year Plan" period. The project of Shikumen building renovation started in 1996, turning its original residential function to new commercial, cultural and public activity functions. In this way, the old Shikumen area with a history of hundreds of years is transformed into a "XinTiandi" (New World), an international community with comprehensive functions.

## 外滩滨水区 | WATERFRONT AREA OF THE BUND

外滩滨水区即外滩建筑群东侧滨江公共空间，北起苏州河口，南至十六铺客运中心北侧边界，总用地面积约 15 公顷。外滩滨水区是上海市最具标志性的城市景观区域，同时也是城市中心最重要的公共活动场所之一。改造前的外滩大部分滨水空间被城市快速机动交通所占用，存在公共活动空间局促、舒适性较差、外滩历史建筑未得到充分展示等问题。为了提升外滩滨水区空间环境品质，迎接世博会的召开，以外滩地下通道的实施为契机，对外滩滨水区域进行综合改造。

外滩滨水区改造致力于体现上海最重要的公共活动空间的特征和现代气息以及外滩地区的历史文化风貌特色，最大限度地为市民提供优美舒适的公共活动空间，打造上海最经典的滨水景观区域和公共活动中心。

**Waterfront Area of the Bund refers to the public spacealong Huangpu River on the east of building complex, stretching from the estuary of Suzhou River in the north to the northern boundary of Shiliupu Passenger Transportation Center in the south and covering about 15 hectares in total. Waterfront Area of the Bund is the core landscape area in Shanghai, and one of the most important public areas in the downtown. Before the reconstruction, the Waterfront Area of the Bund was mostly occupied by fast motor traffic, when limited public area, poor comfortability and inadequate demonstration of historic buildings in the Bund have attracted concerns. In order to improve the urban environment of Waterfront Area of the Bund for the success of the Expo, the area has been comprehensively renovated by virtue of the implementation of underground passage of the Bund.**

**The renovation of Waterfront Area of the Bund stresses the characteristics and modern style of public activity space and the historic and cultural features of the Bund, which are of utmost importance for Shanghai, aiming at minimized the beautiful and comfortable public activity space for citizens and most classic landscape area and public activity center of Shanghai.**

## 延中绿地 | YANZHONG GREEN SPACE

延中绿地位于延安中路高架与南北高架交汇处，占地面积 23 公顷，横跨静安区和黄浦区。

本世纪初，上海的城市绿地建设进入了一个高速发展时期。其中，延中绿地作为市中心大型公共绿地的典型代表，曾是上海旧房密度最高的地区之一，也是上海热岛效应最严重的地区。建成后的延中绿地为改善上海城市面貌、美化城市环境起到了十分重要的作用。

**Yanzhong green space is located in the interchange of Middle Yan'an Road elevated way and the North-South elevated way, covering an area of 23 hectares and involving Jing'an District and Huangpu District.**

**In the beginning of this century, the construction of Shanghai urban green space has entered a period of rapid development. Yanzhong green space, a typical representative of large public green space in the downtown area used to be areas with the highest density of old mansions with most serious heat island effect. After the renovation, Yanzhong green space has been played a very important role in improving the urban landscape and beautifying the urban environment of Shanghai.**

## 尚都里 | SHANGDULI

尚都里位于青浦区朱家角镇的古镇区外围，是一个集旅游、度假、休闲为一体的商业项目。项目沿淀浦河分为南岸和北岸两个片区。已基本建成的南岸片区占地约 3.9 公顷，总建筑规模约 4.7 万平方米，主要由知名的建筑师及建筑设计机构担当设计，空间设计延续了江南水乡古镇的文化特色，而在细节设计中则体现出现代与时尚的韵味。户外空间设计延续了朱家角的江南古镇格局，以街、巷、广场、院、园构成乡村园林的布局。建筑设计融合传统文化元素，有的采用类似"孔明锁"交叉穿插的户型搭接关系，有的以边角通高空间营造飞檐勾角感，并将玻璃金属结合传统材料使用。正在建设的北岸片区的总建筑规模约 6.4 万平方米，改造后片区内原有的粮仓建筑文化将成为一大亮点。尚都里将进一步提升朱家角古镇的旅游休闲吸引力。

Located in the periphery of the ancient town of Zhujiajiao, Qingpu District, Shangduli is a commercial project combining tourism, vacation and leisure together. Along Dianpu River, the project is divided into two areas, namely the area on the southern bank and that on the northern bank. The southern area that is almost finished covers an area about 3.9 hectares with a total building area about 47,000 square meters. Famous architects and four well-known architecture firms are responsible for the design of this area. Its spatial design continues the cultural characteristics of ancient town in the watery region to the south of the Yangtze River, while the detailed design reflects the modern styles and fashion charm. The outdoor spatial design continues the pattern of Zhujiajiao, an ancient town in the watery region to the southern of the Yangtze River. Components such as street, lane, square, courtyard and garden constitute a country garden layout. Architectural design incorporates elements of traditional culture. For example, some buildings in the shape of quasi-"Interlocked" display an overlapping structure. Some buildings with high corners are like angular traditional architecture with cornices. Glass, metal and traditional materials are all used for construction. The northern area under construction has a total building area about 64,000 square meters. After its completion, the granary building that was originally situated in this area will become a major attraction. Shangduli will further enhance the tourism and leisure attraction of Zhujiajiao Ancient Town.

## 世博会城市最佳实践区 | THE URBAN BEST PRACTICES AREA IN THE EXPO

世博会城市最佳实践区位于世博会浦西地区，规划中的文化博览区东侧。世博会后，城市最佳实践区将塑造集创意设计、交流展示、产品体验等为一体，具有世博特征和上海特色的文化创意街区。

借世博会的契机，城市最佳实践区进行了两次更新。第一次更新中，区域布局为“一轴三区”。公共开放空间轴线由南部的全球城市广场、中部的林荫步道和北部的模拟街区广场三部分构成，贯穿南北。第二次更新的布局为“一轴线、两核心、九组团”，整体空间更融为一体。街区的空间形态将延续世博会期间的基本建筑格局，一条步行轴线贯穿整体场地，新老建筑形成紧凑组合，串联开放空间和建筑组团。广场和绿地分别形成南北两个街坊的开放核心。九组建筑组团围绕两个核心，形成功能复合的布局。

另外，对其中的建筑进行重新设计。在整体把控的同时，给予每个建筑改造的空间，形成独特的个性。开放空间系统由广场、步道、院落等形成功能有别、规模不等、形态各异、错落有致、收放相间的连续体系。林荫步道连接南街坊的广场和北街坊的绿地，形成开放空间的主体部分。

The Urban Best Practices Area is located in the Expo Puxi area, the east side of the planning cultural facilities area. After the Expo, the Urban Best Practices Area integrates the creative design, exchange and display together with product experience as a whole, becoming a cultural and creative district with Expo and Shanghai characteristics.

With the opportunity of the Expo, the Urban Best Practices Area has been renovated twice. The layout of the practices area in the first renovation can be summarized as “One Axis, Three Zones.” The public open space axis that runs through north to south of the site consists of three parts, including Global City Square in the south, boulevard in central, and the quasi block plaza in the north. In the second renovation, the layout can be summarized as “One Axis, Two Cores, and Nine Groups.” The overall space is more integrated. The basic building pattern during the Expo retains, with a pedestrian axis running through the whole site. There is a compact combination of old and new buildings to connect open spaces and architectural complex. Open squares and green spaces are the respective core of the northern and southern blocks. Nine building groups center around two cores, forming a layout in line according to the functions.

In addition, there is some architectural redesign to give each building the space of renovation to form a unique personality while considering comprehensive planning. The open space system consists of square, trails and courtyard form a well-arranged continuous system with different functions, varying in size, shapes and forms. The boulevard connects the square in the South district and green land in the North Block to form the main part of the open space.

# 密斯·凡·德·

# SPECIAL EXHIBITION:
# 25 YEARS OF MIES VAN

G

# 罗奖 25 周年纪念展

## DER ROHE AWARD

1988
MADE IN EUROPE
欧洲制造
2015
european union prize for contemporary architecture mies van der rohe award
Mies van der Rohe Foundation, Barcelona

密斯·凡·德·罗奖作为欧洲影响力最大的建筑奖，每两年颁发一次，被认为是欧洲建筑设计作品的最高荣誉。其宗旨为：提供当代建筑与城市规划的交流与探讨的平台，通过传播当代的建筑文化，为建筑教育和认知做出贡献。在当下回溯密斯奖，追问它如何在当时的历史语境中对历史做出评价、判断与反思。本次纪念展将对 20 多年来的获奖作品和优秀作品以文献档案（包括照片、建筑图纸、模型）的形式展出，回顾 1987 年以来欧洲当代的杰出建筑，呈现当代欧洲建筑历史的建构。

**The Mies van der Rohe Award, the most influential architectural prize granted every two years, is acknowledged as the supreme honor for architectural production in Europe. Its goal is to provide a communication platform for contemporary architecture and urban planning, and to contribute to architectural education and understanding through conveying contemporary architectural culture. The exhibition reviews the Award, and how it has criticized, judged, and rethought the history within its specific historical milieu is examined through the collection. This special exhibition archives awarded works since 1987, including photos, architectural drawings, and models, and presents the historical construction of contemporary architecture in Europe.**

ment
PROJECTS
131
TOTAL PROJECTS
2,478
PERCENTAGE (%)
5.29
Mixed use
146
2,478
5.89
Commerce
PROJECTS
128
TOTAL PROJECTS
2,478
PERCENTAGE (%)
5.17
05 2009
07 2009
2009

**密斯·凡·德·罗基金会，巴塞罗那 | Fundació Mies van der Rohe, Barcelona**

1983 年，为了重建 1929 年密斯·凡·德·罗设计的世界博览会德国馆，密斯凡德罗基金会创建。自 1988 年以来，在欧洲联盟的支持下，基金会组织了欧洲联盟—当代建筑奖。过去 25 年间，密斯·凡·德·罗奖已逐渐成为欧洲建筑界的最高荣誉奖项。 In 1983 the Mies van der Rohe Foundation was created with the aim of reconstructing the German Pavilion that Ludwig Mies van der Rohe created for the 1929 Universal Exhibition. Since 1988 it has organised, with the support of the European Union, the European Union Prize for Contemporary Architecture – Mies van der Rohe Award which over the last 25 years has become the leading reference for architecture on the continent.

2015年9月29日，2015上海城市空间艺术季主展览在西岸艺术中心开幕。2015城市空间艺术季主展览中方总策展人伍江，外方总策展人莫森·莫斯塔法维，本届艺术季学术委员会主任、中国科学院院士、法国建筑科学院院士郑时龄，本届艺术季学术委员会主任、中国美术家协会副主席、中国美术馆馆长、全国城市雕塑建设指导委员会副主任吴为山，上海市人民政府副秘书长黄融，先后为开幕式致辞，分享展览主题"城市更新"的解读及对2015城市空间艺术季活动的寄望。上海戏剧学院和华东师范大学的学生为观众带来《城市漫步》《灵魂的舞动》和《生命的飞跃》的主题舞蹈。

On September 29, 2015, the 2015 SUSAS main exhibition was opened in West Coast Art Center. China's chief curator of the 2015 SUSAS main exhibition Wu Jiang, foreign chief curator Mohsen Mostafavi, the current Artistic Director of the Academic Committee, CAS member and the academician of French Academy of Architecture Sciences Zheng Shiling, the Current Artistic Director of the Academic Committee, Vice-chairman of China Artists Association, Director of China National Gallery of Art and Deputy Director of the Steering Committee of the National Urban Sculpture Construction Wu Weishan and Deputy Secretary of Shanghai Municipal People's Government Huang Rong successively addressed the opening ceremony, and shared the interpretation of the exhibition theme "Urban Renewal" as well as their expectations for the 2015 SUSAS activity

Students from Shanghai Theater Academy and East China Normal University performed theme dances including City Stroll, Movement from the Soul and The Leap of Life.

SUSAS
上海城市空间艺术季

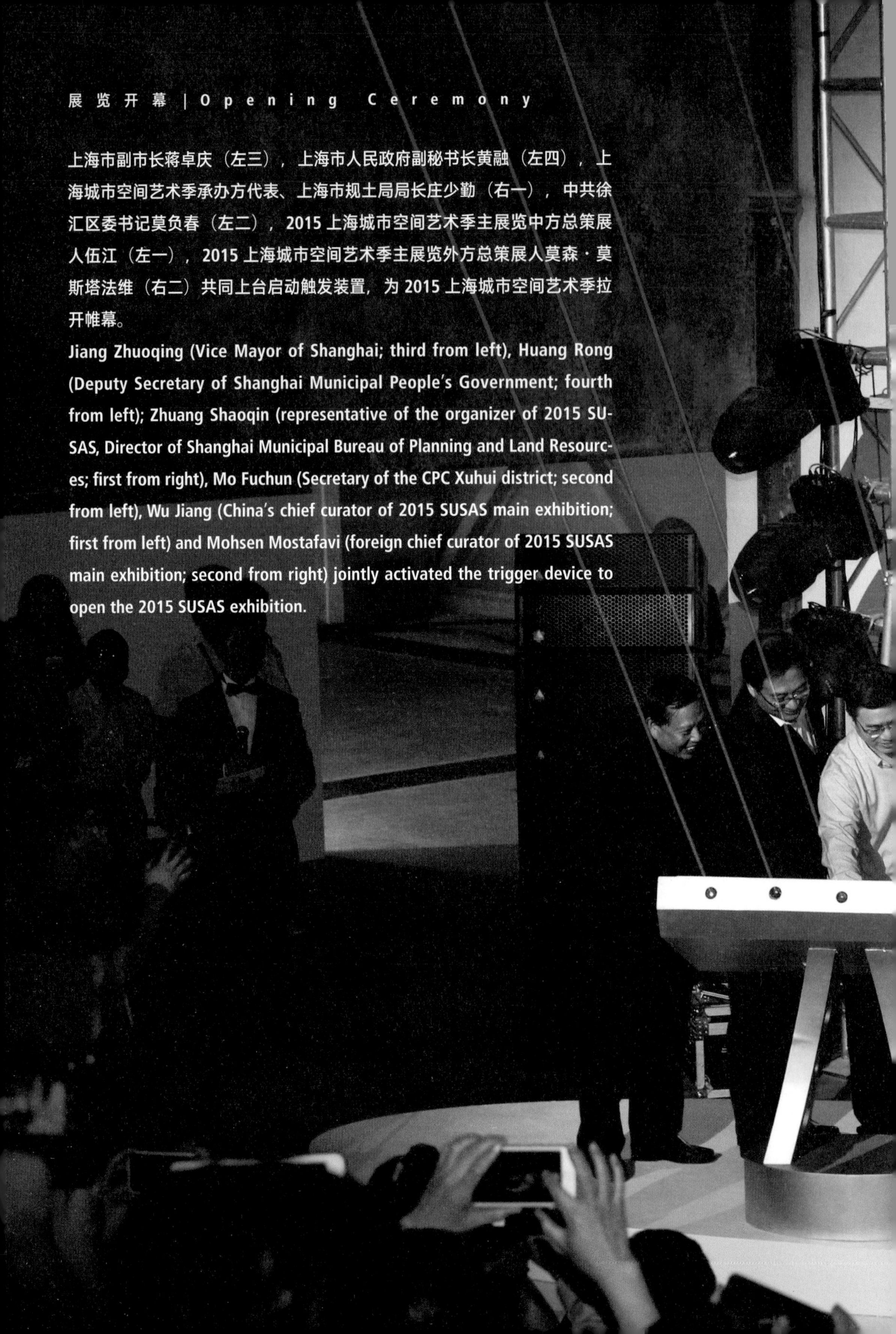

上海市副市长蒋卓庆（左三），上海市人民政府副秘书长黄融（左四），上海城市空间艺术季承办方代表、上海市规土局局长庄少勤（右一），中共徐汇区委书记莫负春（左二），2015 上海城市空间艺术季主展览中方总策展人伍江（左一），2015 上海城市空间艺术季主展览外方总策展人莫森·莫斯塔法维（右二）共同上台启动触发装置，为 2015 上海城市空间艺术季拉开帷幕。

Jiang Zhuoqing (Vice Mayor of Shanghai; third from left), Huang Rong (Deputy Secretary of Shanghai Municipal People's Government; fourth from left); Zhuang Shaoqin (representative of the organizer of 2015 SUSAS, Director of Shanghai Municipal Bureau of Planning and Land Resources; first from right), Mo Fuchun (Secretary of the CPC Xuhui district; second from left), Wu Jiang (China's chief curator of 2015 SUSAS main exhibition; first from left) and Mohsen Mostafavi (foreign chief curator of 2015 SUSAS main exhibition; second from right) jointly activated the trigger device to open the 2015 SUSAS exhibition.

SUSAS

住房和城乡建设部参观

联合国人居属参观

江苏省住房和城乡建设厅参观

广东省住房和城乡建设厅参观

成都市规划管理局参观

遵义市城乡规划局参观

上海市静安区规土局参观

上海市宝山区规土局参观

中国民主促进会上海市委参观

商务部国土部合办发展中国家土地利用规划研修班参观

上海规划院发展研究中心参观

世博博物馆馆长参观

九三学社参观

市青年联合会参观

同济校友参观

名苑小区居委参观

澎湃新闻参观

总策展人　伍江

总策展人　莫森 · 莫斯塔法维

郑时龄院士在开幕现场作主题演讲

吴为山教授在开幕现场作主题演讲

日本大地艺术发起人、策展人北川弗兰作主题报告

新闻发布会

城市－建筑策展人李翔宁教授现场导览

上海特展策展人章明教授现场导览

《城影相间》摄影展开幕式

INTERIOR 建筑内向展论坛

上海市青少年发展“十三五”规划青年汇智团“汇智讲堂”讲座

碧山建造学社暨同绿分享会工艺沙龙

创建竹艺的传统

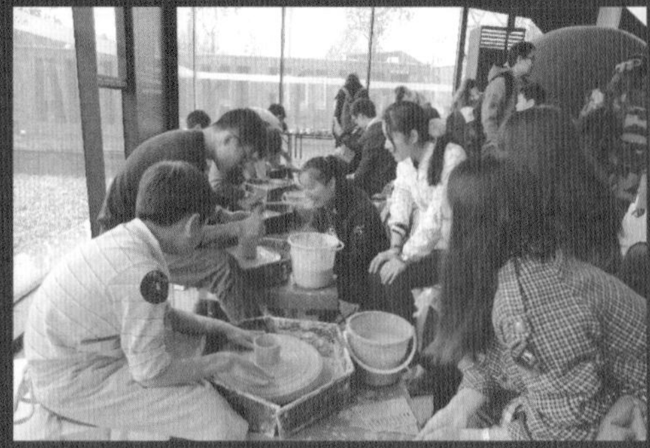

火热秀场、非泥不可活动现场

微信会员散客拼团活动

小小建筑师工作坊

西班牙美食周

小小建筑师公开课

弘虹艺术机构团体参观

“互联网与社会”论坛

弘虹艺术机构儿童写生

2015 国际建筑评论家委员会上海研讨会嘉宾讨论

王婷婷老师绘画

2015 国际建筑评论家委员会上海研讨会 演讲嘉宾矶崎新与主持人李翔宁

售票
Tickets
寄包
Cloakroom
ART WEST BU

西岸
WEST BUND ART CENTER

**西岸艺术中心**
**WEST BUND ART CENTER**

**建筑师：大舍建筑设计事务所**
**设计团队：柳亦春、王龙海、伍正辉、王伟实**
**结构与机电工程师：同济大学建筑设计研究院（集团）有限公司**
**地址：上海市徐汇区龙腾大道 2555 号**
**建筑面积：10 800 平方米**
**设计时间：2014 年 3 月至 7 月**
**建成时间：2014 年 9 月**

**Architect: Atelier Deshaus**
**Design team: Liu Yichun, Wang Longhai, Wu Zhenghui, Wang Weishi**
**Structural, electrical and mechanical engineer: Tongji University Architectural Design & Research Institute (Group) Co., Ltd.**
**Address: 2555 Longteng Avenue, Xuhui District, Shanghai**
**Building Area: 10,800 m$^2$**
**Design period: March to July 2014**
**Time of completion: September 2014**

西岸艺术中心是由上海飞机制造厂冲压车间的厂房改造而成，这个长达 120 米、宽度两跨 24 米和 30 米、高度 15 米的巨大厂房，这个巨大覆盖的单个空间，给人强烈的安全感和自由感，这在通常的民用建筑尺度下是非常少见的，所以维持这个空间的完整性与尺度感有着特殊的意义。

因为西侧城市道路的原因，厂房西段有四跨被拆除，在恢复抗震柱的同时，设计采用了菱形网状的钢结构网格加强了建筑的侧向稳定性，建筑借机向城市呈现一定的开放性，西侧是从地铁方向人流到达的方向，一个新的立面，特别是傍晚灯光亮起时如灯笼般的空透，使这个旧有的厂房建立了和周边正在重建更新的城市间的关联。

出于功能的考虑，24 米跨这一部分的厂房内设置了夹层，分别通过阶梯看台和一个折行坡道与 30 米跨的厂房空间形成了闭合的观展流线，于是，一个从夹层高处望向东侧黄浦江的视线需求出现了。厂房东侧的山墙被部分打开，并采用了新的混凝土构架和旧有的抗震柱以及圈梁连结为新的格构，弯折的坡道配合着不同高度不同方向的视线联系着龙腾大道和黄浦江的方向，这里，是太阳升起的方向。而完整保留了原有尺度的 30 米跨的大空间，在朝向城市这一侧，菱形的网格后，每每有红日西沉。这里，被许多人称为“日落大厅”，西岸的落日逐渐变成了这个城市的一道闪亮的风景。

2015 上海城市空间艺术季主展场所在地——上海西岸，曾是近代海派文化的发源地、中西文化的交汇地和现代文化产业的诞生地，见证了上海乃至整个中国现代工业发展的历史。以“西岸文化走廊”品牌工程、“西岸传媒港”等项目为核心，西岸目前不仅创建了西岸音乐节、西岸建筑与当代艺术双年展、西岸艺术与设计博览会等品牌活动，还成功吸引了龙美术馆、余德耀美术馆、上海梦中心、东方梦工厂、上海摄影艺术中心等项目的入驻，旨在打造与巴黎左岸、伦敦南岸比肩的独具魅力的城市文化新地标。

West Bund Art Center is adapted from the stamping workshop of Shanghai Aircraft Manufacturing Co., Ltd. The huge workshop has a length of 120 meters, spans of 24 meters and 30 meters, and a height of 15 meters. This single structure with a large area delivers a strong sense of safety and freedom, which is rarely seen in the common civil architecture scale. Therefore, it's of special significance to maintain the integrity and special scale of this structure.

Four piers on the west section of the workshop are demolished due to the city road in the west side. Earthquake-resistant columns are recovered; at the same time, rhombic and reticulate grids of steel structure are adopted to enhance the lateral stability of the structure. The structure is connected to the city with its west side leading to the metro station where there are a large number of people. This new façade looks like a lantern in the lamplight at dusk; it connects the old workshop with the surrounding city under reconstruction and renewal.

In consideration of function, an interlayer is designed in the 24-meter-span section of the workshop. With the terrace and zigzag ramp, a closed visiting streamline is formed between this section and the 30-meter-span section. Visitors may need to see the Huangpu River in the east at the top of the interlayer, so part of the gable on the east side of the workshop is demolished. New concrete frames, existing earthquake-resistant columns and ring beams are joined together to build a new lattice. The zigzag ramp in different heights and directions delivers a view of Longteng Avenue and Huangpu River where the sun rises. The original 30-meter-span space is kept intact. On the side facing the city and behind the rhombic grids, the sun sets in the west. This is called "Sunset Hall" by many. The setting sun of the West Bund gradually becomes a fascinating view in the city.

The site that the main exhibition of SUSAS 2015 located in distinguished as a historic cradle of Shanghai's regional culture and an important convergence point for Chinese and Western cultures. Later on, it also played a central role in the development of China's modern cultural industry, bearing witness to the progress of modern industry in Shanghai – as well as the nation. As important elements of two projects focused on the area's cultural development, the West Bund Cultural Corridor Construction project and the West Bund Media Port project, the West Bund hosts a diverse range of branded events including the Shanghai West Bund Music Festival, the West Bund Architecture and Contemporary Art Biennial and the West Bund Art & Design Fair. The West Bund has also attracted an array of renowned cultural organizations to move in, including the Long Museum (West Bund), the Yuz Museum, the Shanghai Dream Center, the Oriental DreamWorks and the Shanghai Center of Photography. Ultimately, the area aims to become a cultural landmark in Shanghai, on par with La Rive Gauche in Paris and the South Bank in London.

| 日期 | 地点 | 论坛 / 活动 |
| --- | --- | --- |
| 9/29 | | 2015 上海城市空间艺术季主题论坛<br>+ 圆桌论坛 |
| | | 新闻发布会 |
| | | 学委会学术论坛 |
| | | 主展览开幕仪式 |
| 9/30 | | “城市更新”学术论坛 |
| 10/17 | C10 展位二楼 | 碧山建造学社暨同绿分享会工艺沙龙<br>第一回：水磨石工艺 |
| 10/24 | 艺术示范区 - 周铁海工作室 | 2015《城影相间》摄影展开幕式 |
| 10/25 | 大台阶 | 上海市青少年发展“十三五”规划<br>青年汇智团“汇智讲堂”讲座 |
| | 大台阶 | INTERIOR 西班牙建筑内向展论坛 |
| 10/31 | 旋转台阶 | 城市音乐空间表演 |
| | 主展馆 | 弘虹艺术机构美术写生一 |
| | C-10 展位二楼 | 碧山建造学社暨同绿分享会工艺沙龙<br>第二回：“清水混凝土修补及模板工艺” |

| | | |
|---|---|---|
| **11/1** | 旋转台阶 | 城市音乐空间表演 |
| | 主展馆 | 弘虹艺术机构美术写生二 |
| **11/7** | 旋转台阶 | 城市音乐空间表演 |
| **11/8** | C10 展位二楼 | 碧山建造学社暨同绿分享会工艺沙龙<br>第三回：“漆语——传统大漆工艺与漆画工艺” |
| **11/14** | 旋转台阶 | 城市音乐空间表演 |
| **11/15** | C6 展位 | 陈浩如竹构展位讲座：<br>创建竹艺的传统 + 合力搭建纪实 |
| | C10 展位二楼 | 碧山建造学社暨同绿分享会工艺沙龙<br>第四回：“文人家具系统与工艺” |
| | C10 展位二楼 | 木作经典之隆兴寺的转轮藏讲座 |
| | 旋转台阶 | 城市音乐空间表演 |
| | | “散客拼团”来看展 |
| **11/21** | 咖啡厅 | 火热 SHOW 场灯工玻璃<br>+ 非泥不可 DIY 陶艺手工坊 |
| | | “散客拼团”来看展 |
| | 旋转台阶 | 城市音乐空间表演 |
| **11/22** | 咖啡厅 | 火热 SHOW 场灯工玻璃<br>+ 非泥不可 DIY 陶艺手工坊 |
| | 二楼上海特展展区内 | 西班牙美食周活动开幕 |
| **11/28** | 旋转台阶 | 城市音乐空间表演 |
| **11/29** | 大台阶 | 小小建筑师公开课 |
| | 咖啡厅 | 小小建筑师工作坊——理想的住宅 |
| | 旋转台阶 | 城市音乐空间表演 |
| **12/5** | 大台阶 | 互联网与社会讲座 |
| | 旋转台阶 | 城市音乐空间表演 |
| **12/6** | 旋转台阶 | 城市音乐空间表演 |
| **12/11** | 报告厅 | 2015 国际建筑评论家委员会上海研讨会：<br>超越东西南北——当代建筑的普遍性与特殊性 |
| **12/12** | 旋转台阶 | 城市音乐空间表演 |
| **12/13** | 旋转台阶 | 城市音乐空间表演 |

# 致谢

经过一年多的筹备，2015 上海城市空间艺术季主展览在策展团队、国内外学者，以及社会各界的共同努力下成功举办了。来自世界各地的规划师、建筑师和艺术家们通过主题为"城市更新"的主展览，展示了各自多年来的研究成果，为上海的城市建设、城市更新工作提出了独到的见解，同时也为上海市民搭建了一个了解城市更新历史、感受上海城市文化的互动平台。展览期间，社会各界对此次展览的内容、形式以及布展方式等均给予了高度的评价，同时也提出了很多有建设性的意见和建议。此次展览的举办为组织下一届上海城市空间艺术季积累了宝贵的工作经验。

展览在举办过程中得到了多方的大力支持和帮助，谨代表 2015 上海城市空间艺术季的承办单位，特别鸣谢以下单位和团体：浦东新区人民政府、黄浦区人民政府、静安区人民政府、长宁区人民政府、普陀区人民政府、闸北区人民政府（原）、虹口区人民政府、杨浦区人民政府、宝山区人民政府、闵行区人民政府、嘉定区人民政府、金山区人民政府、松江区人民政府、青浦区人民政府、奉贤区人民政府、崇明县人民政府、上海市城市规划设计研究院、上海西岸开发（集团）有限公司、上海市测绘院。另外，我们也要感谢对艺术季曾经给予支持和付出努力的领导、专家、策展人、工作人员和热情参与的市民。

此外，本画册中部分资料和图片由原创作者免费提供，在此一并表示感谢！

**上海市规划和国土资源管理局**
**上海市文化广播影视管理局**
**上海市徐汇区人民政府**

# ACKNOWLEDGEMENTS

After more than a year of preparation, the main exhibition of Shanghai Urban Space Art Season 2015 was successfully held with the joint efforts of the curatorial team, domestic and international scholars and people from all walks of life. At the main exhibition themed by "Urban Regeneration", planners, architects and artists from all over the world displayed their achievements of many years' research, offering unique insights into urban construction and regeneration in Shanghai and creating for Shanghai residents an interactive platform to understand the history of urban regeneration and feeling the urban culture of Shanghai. During the exhibition, people from all sectors of society spoke highly of the content, form and layout methods of the show while providing many constructive comments and suggestions. From this exhibition, we have gained valuable experience for the next Shanghai Urban Space Art Season.

We have gained vigorous support and assistance from multiple parties during the exhibition. Hereby on behalf of the organizer of Shanghai Urban Space Art Season 2015, we would like to express our special gratitude to the following entities and organizations:

People's Government of Pudong New Area, People's Government of Huangpu District, People's Government of Jing'an District, People's Government of Changning District, People's Government of Putuo District, People's Government of Zhabei District (former), People's Government of Hongkou District, People's Government of Yangpu District, People's Government of Baoshan District, People's Government of Minhang District, People's Government of Jiading District, People's Government of Jinshan District, People's Government of Songjiang District, People's Government of Qingpu District, People's Government of Fengxian District, People's Government of Chongming County, Shanghai Urban Planning and Design Research Institute, Shanghai West Bund Development (Group) Co. Ltd. and Shanghai Municipal Institute of Surveying and Mapping. Meanwhile, we would like to appreciate those people who have given support and made efforts on SUSAS including experts, curators, staff, citizens and their enthusiastic participation.

In addition, some data and pictures in this publication are provided by courtesy of the authorship. Thank you for all the support!

**Shanghai Municipal Bureau of Planning and Land Resources**
**Shanghai Municipal Administration of Culture, Radio, Film & TV**
**the People's Government of Xuhui District**

2015 上海城市空间艺术地图
SHANGHAI URBAN SPACE ART MAP
图例 LEGENDS
主展览
MAIN EXHIBITION PAVILION
展览标志性建筑
LANDMARK OF EXHIBITION
案例展
SITE PROJECTS
主要道路
MAIN ROAD
联合展
ANCILLARY EXHIBITIONS
铁路及站名
RAILWAY & STATION
部分上海优秀城市空间案例
SOME OUTSTANDING CASES OF SHANGHAI URBAN SPACE
轨道交通、车站及线路序号
METROLINE, METRO STATION, METRO NUMBER
高速公路及出入口
EXPRESSWAY ENTRANCE-EXIT
隧道
TUNNEL
高架路、城市快速路及匝道
ELEVATED ROAD, FREEWAY, RAMP
轮渡线
FERRY LINE
杨家桥 Yangjiaqiao
上海西站 Shanghai West Railway Station
真如 Zhenru
陆家宅 Lujiazhai
曹家渡 Caojiadu
忻康里 Xinkangli
北新泾 Beixinjing
周家桥 Zhouqiaqiao
中山公园 Zhongshan Park
静安寺 Jing'An Temple
上海动物园 Shanghai Zoo
徐家汇 Xujiahui
龙华 Longhua
漕河泾 Caohejing
新龙华 Xinlonghua
上海南站 Shanghai South Railway Station
大宁灵石公园 Daning Lingshi Park
长风公园 Changfeng Park

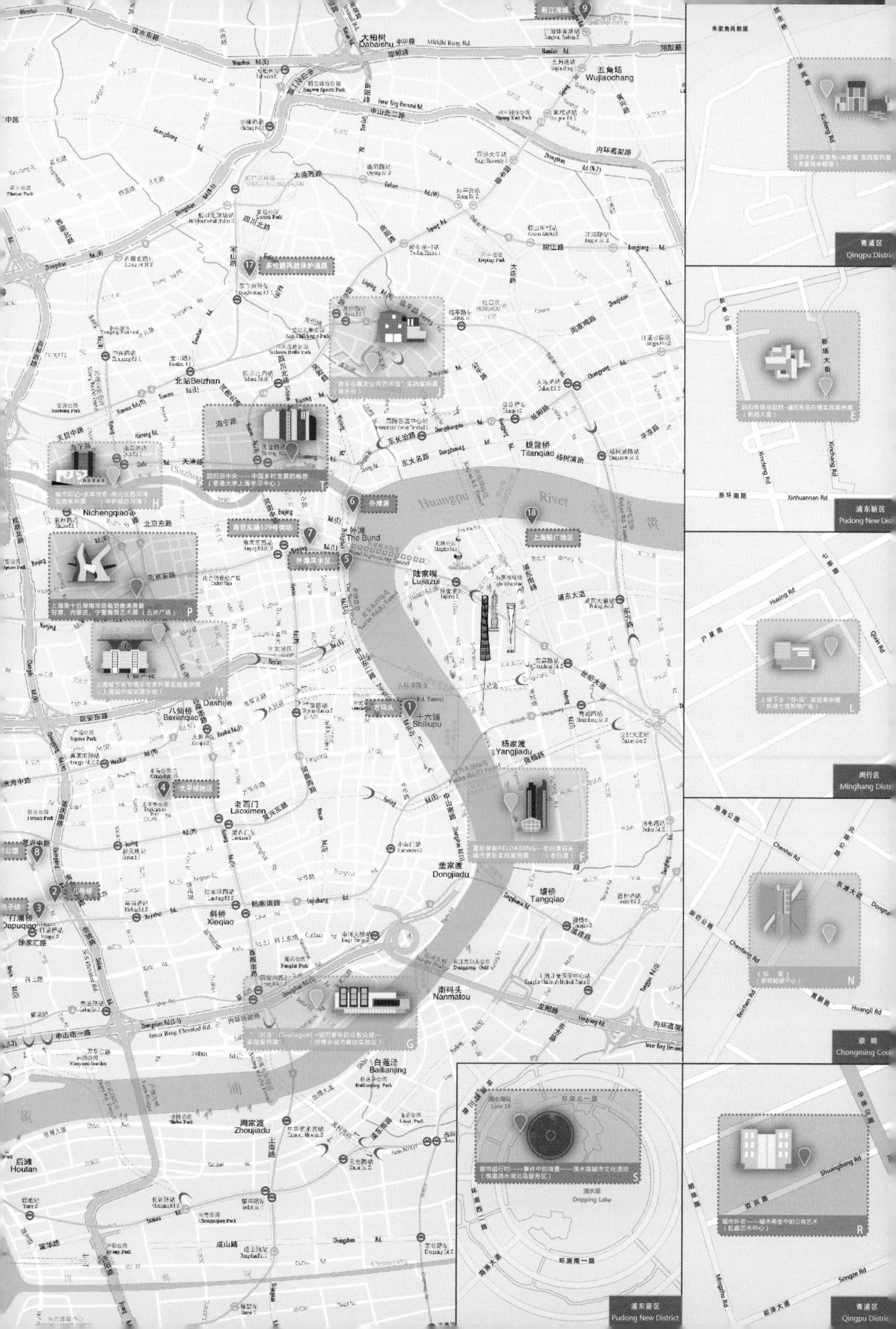

大柏树 Dabaishu
五角场 Wujiaochang
9 新江湾城
17 多伦路风貌保护道路
北站 Beizhan
Nichengqiao
北京东路
南京东路179号街坊
7
6 外滩源
5 外滩滨水区
外滩 The Bund
Huangpu River
18 上海船厂地区
陆家嘴 Lujiazui
提篮桥 Tilanqiao
东长治路
东大名路
天目中路
天潼路
海宁路
南京东路
延安东路
八仙桥 Baxianqiao
Dashijie
1 老码头
十六铺 Shiliupu
杨家渡 Yangjiadu
4 大平桥地区
老西门 Laoxīmen
8
2 八号桥
3
打浦桥 Dapuqiao
徐家汇路
斜桥 Xieqiao
董家渡 Dongjiadu
塘桥 Tangqiao
南码头 Nanmatou
白莲泾 Bailianjing
周家渡 Zhoujiadu
后滩 Houtan
中山南一路
中山北二路
内环高架路
浦东大道
世纪大道
东方路
滴水湖 Dripping Lake
环湖北一路
环湖南一路
Huxing Rd
Qixin Rd
Chenhai Rd
Chenfang Rd
Huangli Rd
Shuangbang Rd
Songze Rd
Xinfeng Rd
Xinchang Rd
Xinhuannan Rd
H
T
J
P
M
F
G
S
E
L
N
R
青浦区 Qingpu District
浦东新区 Pudong New District
闵行区 Minghang District
崇明 Chongming County

图书在版编目（CIP）数据

2015 上海城市空间艺术季主展览 / 上海城市空间艺

术季展览画册编委会编 . -- 上海 : 同济大学出版社 ,

2016.7

ISBN 978-7-5608-6109-8

Ⅰ . ① 2... Ⅱ . ①上 ... Ⅲ . ①城市规划一空间规划一

上海市一画册 Ⅳ . ① TU984.251-64

中国版本图书馆 CIP 数据核字 (2015) 第 297223 号

2015 上海城市空间艺术季主展览
*2015 Shanghai Urban Space Art Season*
*Main Exhibition*

上海城市空间艺术季展览画册编委会 / 编
Edited by SUSAS Publication Editorial Board

策划：上海城市公共空间设计促进中心 群岛工作室
项目统筹：马宏
责任编辑： 杨碧琼
特约编辑：邹野
责任校对： 徐春莲
封面设计：绵延工作室 | Atelier Mio
装帧设计： typo_d
展场摄影：蒋祖龙（除特别标注外）
版 次：2016 年 7 月第 1 版
印 次：2016 年 7 月第 1 次印刷
印 刷：上海安兴汇东纸业有限公司
开 本：787mm × 1092mm 1/16
印 张：21.75
字 数：435 000
ISBN：978-7-5608-6109-8
定 价：198.00 元
出版发行：同济大学出版社
地 址：上海市四平路 1239 号
邮政编码：200092
网 址：http://www.tongjipress.com.cn
经 销：全国各地新华书店